Lecture Notes in Mathematics

For information about Vols. 1–342, please contact your bookseller or Springer-Verlag.

Vol. 343: Algebraic K-Theory III, Hermitian K-Theory and Geometric Applications. Edited by H. Bass. XV, 572 pages. 1973.

Vol. 344: A. S. Troelstra (Editor), Metamathematical Investigation of Intuitionistic Arithmetic and Analysis. XVII, 485 pages. 1973.

Vol. 345: Proceedings of a Conference on Operator Theory. Edited by P. A. Fillmore. VI, 228 pages. 1973.

Vol. 346: Fučík et al., Spectral Analysis of Nonlinear Operators. II, 287 pages. 1973.

Vol. 347: J. M. Boardman and R. M. Vogt, Homotopy Invariant Algebraic Structures on Topological Spaces. X, 257 pages. 1973.

Vol. 348: A. M. Mathai and R. K. Saxena, Generalized Hypergeometric Functions with Applications in Statistics and Physical Sciences. VII, 314 pages. 1973.

Vol. 349: Modular Functions of One Variable II. Edited by W. Kuyk and P. Deligne. V, 598 pages. 1973.

Vol. 350: Modular Functions of One Variable III. Edited by W. Kuyk and J.-P. Serre. V, 350 pages. 1973.

Vol. 351: H. Tachikawa, Quasi-Frobenius Rings and Generalizations. XI, 172 pages. 1973.

Vol. 352: J. D. Fay, Theta Functions on Riemann Surfaces. V, 137 pages. 1973.

Vol. 353: Proceedings of the Conference on Orders, Group Rings and Related Topics. Organized by J. S. Hsia, M. L. Madan and T. G. Ralley. X, 224 pages. 1973.

Vol. 354: K. J. Devlin, Aspects of Constructibility. XII, 240 pages. 1973.

Vol. 355: M. Sion, A Theory of Semigroup Valued Measures. V, 140 pages. 1973.

Vol. 356: W. L. J. van der Kallen, Infinitesimally Central Extensions of Chevalley Groups. VII, 147 pages. 1973.

Vol. 357: W. Borho, P. Gabriel und R. Rentschler, Primideale in Einhüllenden auflösbarer Lie-Algebren. V, 182 Seiten. 1973.

Vol. 358: F. L. Williams, Tensor Products of Principal Series Representations. VI, 132 pages. 1973.

Vol. 359: U. Stammbach, Homology in Group Theory. VIII, 183 pages. 1973.

Vol. 360: W. J. Padgett and R. L. Taylor, Laws of Large Numbers for Normed Linear Spaces and Certain Fréchet Spaces. VI, 111 pages. 1973.

Vol. 361: J. W. Schutz, Foundations of Special Relativity: Kinematic Axioms for Minkowski Space-Time. XX, 314 pages. 1973.

Vol. 362: Proceedings of the Conference on Numerical Solution of Ordinary Differential Equations. Edited by D.G. Bettis. VIII, 490 pages. 1974.

Vol. 363: Conference on the Numerical Solution of Differential Equations. Edited by G. A. Watson. IX, 221 pages. 1974.

Vol. 364: Proceedings on Infinite Dimensional Holomorphy. Edited by T. L. Hayden and T. J. Suffridge. VII, 212 pages. 1974.

Vol. 365: R. P. Gilbert, Constructive Methods for Elliptic Equations. VII, 397 pages. 1974.

Vol. 366: R. Steinberg, Conjugacy Classes in Algebraic Groups (Notes by V. V. Deodhar). VI, 159 pages. 1974.

Vol. 367: K. Langmann und W. Lütkebohmert, Cousinverteilungen und Fortsetzungssätze. VI, 151 Seiten. 1974.

Vol. 368: R. J. Milgram, Unstable Homotopy from the Stable Point of View. V, 109 pages. 1974.

Vol. 369: Victoria Symposium on Nonstandard Analysis. Edited by A. Hurd and P. Loeb. XVIII, 339 pages. 1974.

Vol. 370: B. Mazur and W. Messing, Universal Extensions and One Dimensional Crystalline Cohomology. VII, 134 pages. 1974.

Vol. 371: V. Poenaru, Analyse Différentielle. V, 228 pages. 1974.

Vol. 372: Proceedings of the Second International Conference on the Theory of Groups 1973. Edited by M. F. Newman. VII, 740 pages. 1974.

Vol. 373: A. E. R. Woodcock and T. Poston, A Geometrical Study of the Elementary Catastrophes. V, 257 pages. 1974.

Vol. 374: S. Yamamuro, Differential Calculus in Topological Linear Spaces. IV, 179 pages. 1974.

Vol. 375: Topology Conference. Edited by R. F. Dickman Jr. and P. Fletcher. X, 283 pages 1974.

Vol. 376: I. J. Good and D. B. Osteyee, Information, Weight of Evidence. The Singularity between Probability Measures and Signal Detection. XI, 156 pages. 1974.

Vol. 377: A. M. Fink, Almost Periodic Differential Equations. VIII, 336 pages. 1974.

Vol. 378: TOPO 72 – General Topology and its Applications. Proceedings 1972. Edited by R. A. Alò, R. W. Heath and J. Nagata. XIV, 651 pages. 1974.

Vol. 379: A. Badrikian et S. Chevet, Mesures Cylindriques, Espaces de Wiener et Fonctions Aléatoires Gaussiennes. X, 383 pages. 1974.

Vol. 380: M. Petrich, Rings and Semigroups. VIII, 182 pages. 1974.

Vol. 381: Séminaire de Probabilités VIII. Edité par P. A. Meyer. IX, 354 pages. 1974.

Vol. 382: J. H. van Lint, Combinatorial Theory Seminar Eindhoven University of Technology. VI, 131 pages. 1974.

Vol. 383: Séminaire Bourbaki – vol. 1972/73. Exposés 418–435. IV, 334 pages. 1974.

Vol. 384: Functional Analysis and Applications, Proceedings 1972. Edited by L. Nachbin. V, 270 pages. 1974.

Vol. 385: J. Douglas Jr. and T. Dupont, Collocation Methods for Parabolic Equations in a Single Space Variable (Based on C^1-Piecewise-Polynomial Spaces). V, 147 pages. 1974.

Vol. 386: J. Tits, Buildings of Spherical Type and Finite BN-Pairs. X, 299 pages. 1974.

Vol. 387: C. P. Bruter, Eléments de la Théorie des Matroïdes. V, 138 pages. 1974.

Vol. 388: R. L. Lipsman, Group Representations. X, 166 pages. 1974.

Vol. 389: M.-A. Knus et M. Ojanguren, Théorie de la Descente et Algèbres d' Azumaya. IV, 163 pages. 1974.

Vol. 390: P. A. Meyer, P. Priouret et F. Spitzer, Ecole d'Eté de Probabilités de Saint-Flour III – 1973. Edité par A. Badrikian et P.-L. Hennequin. VIII, 189 pages. 1974.

Vol. 391: J. W. Gray, Formal Category Theory: Adjointness for 2-Categories. XII, 282 pages. 1974.

Vol. 392: Géométrie Différentielle, Colloque, Santiago de Compostela, Espagne 1972. Edité par E. Vidal. VI, 225 pages. 1974.

Vol. 393: G. Wassermann, Stability of Unfoldings. IX, 164 pages. 1974.

Vol. 394: W. M. Patterson, 3rd, Iterative Methods for the Solution of a Linear Operator Equation in Hilbert Space – A Survey. III, 183 pages. 1974.

Vol. 395: Numerische Behandlung nichtlinearer Integrodifferential- und Differentialgleichungen. Tagung 1973. Herausgegeben von R. Ansorge und W. Törnig. VII, 313 Seiten. 1974.

Vol. 396: K. H. Hofmann, M. Mislove and A. Stralka, The Pontryagin Duality of Compact O-Dimensional Semilattices and its Applications. XVI, 122 pages. 1974.

Vol. 397: T. Yamada, The Schur Subgroup of the Brauer Group. V, 159 pages. 1974.

Vol. 398: Théories de l'Information, Actes des Rencontres de Marseille-Luminy, 1973. Edité par J. Kampé de Fériet et C.-F. Picard. XII, 201 pages. 1974.

continuation on page 185

Lecture Notes in Mathematics

Edited by A. Dold and B. Eckmann

572

Sparse Matrix Techniques

Copenhagen 1976

Advanced Course Held at the Technical University of Denmark Copenhagen, August 9-12, 1976

Edited by V. A. Barker

Springer-Verlag
Berlin Heidelberg GmbH 1977

Editor
V. A. Barker
Institute for Numerical Analysis
Technical University of Denmark
Copenhagen/Denmark

Library of Congress Cataloging in Publication Data

Course in Advanced Sparse Matrix Techniques, Technical University of Denmark, 1976.
Sparse matrix techniques.

(Lecture notes in mathematics ; 572)
1. Matrices--Addresses, essays, lectures. 2. Equations, Simultaneous--Addresses, essays, lectures.
I. Barker, Vincent Allan, 1934- II. Title.
III. Series: Lecture notes in mathematics (Berlin) ; 572.
QA3.L28 no. 572 [QA185] 510'.8s [512.9'43] 77-1128

AMS Subject Classifications (1970): 65-02, 65F05, 65F10, 65F15, 65F20, 65N20, 94A20

ISBN 978-3-540-08130-2 ISBN 978-3-540-37430-5 (eBook)
DOI 10.1007/978-3-540-37430-5

Originally published by Springer-Verlag Berlin Heidelberg New York in 1977

PREFACE

These notes were written for the "Course in Advanced Sparse Matrix Techniques" held at the Technical University of Denmark, Copenhagen, on August 9-12, 1976. The course was arranged by the Institute for Numerical Analysis, a department of the University, with the financial support of the Danish Natural Science Research Council. There were 72 participants from 9 countries.

The purpose of the course was to present scientists and engineers in higher education and industry with state-of-the-art material in one of the most rapidly growing areas in numerical analysis. It was decided at the outset to concentrate on a few topics of central importance rather than attempt to be comprehensive, and to invite as lecturers leading researchers in those topics. The program was the following:

OWE AXELSSON, Chalmers University of Technology, Sweden;
Solution of linear systems of equations: iterative methods.

J. ALAN GEORGE, University of Waterloo, Canada;
Solution of linear systems of equations: direct methods for finite element problems.

JOHN K. REID, Atomic Energy Research Establishment, Harwell, England;
Solution of linear systems of equations: direct methods (general).

AXEL RUHE, University of Umeå, Sweden;
Computation of eigenvalues and eigenvectors.

Four 1-hour lectures were devoted to each topic.

Most of the staff of the Institute for Numerical Analysis contributed in one way or another to this project, and it is a pleasure to acknowledge their support. Thanks are due in particular to Karen Margrethe Arildsen, Thøger Busk and Ole Tingleff for their part in the organizational effort, and to Inger Margrethe Kruber for her assistance in the editing of these notes. Most of all, we must thank the 4 lecturers who both on and off the lecture floor so ably shared with us their profound knowledge of this fascinating subject.

V. A. Barker
Copenhagen, 1976

TABLE OF CONTENTS

SOLUTION OF LINEAR SYSTEMS OF EQUATIONS: ITERATIVE METHODS

Owe Axelsson

Dept. of Computer Sciences
Chalmers University of Technology
S 402 20 Gothenburg, Sweden

CONTENTS

1. Introduction.

This lecture series will deal with sparse matrix techniques. For years (see Section 7.1), iterative methods have been used for the solution of large linear systems of equations. The obvious advantage with iterative methods is that they usually demand a minimum of storage: only non-zero coefficients of the given matrix have to be stored plus the coefficients in one or a few vectors such as the solution vector and, for example, the residual vector.

For certain classes of matrices, like diagonally dominant matrices, the simplest methods such as the simultaneous iteration method and the successive overrelaxation (SOR) method, with overrelaxation parameter $0 < \omega < 2$, are convergent (see e.g. [1]). For faster convergence, the class of matrices to be dealt with must possess further properties.

For instance, a classical result for consistently ordered matrices (2-cyclic matrices, see [1]) is that the convergence of the SOR method with optimal parameter ω is given by the spectral radius $\rho(\mathcal{L}_\omega) = 2/[1 + \sqrt{1-\rho(B)^2}] - 1$ (cf. Section 5.4).

It is apparent that the more properties the class of matrices at hand satisfies, the faster are the methods that can be developed taking advantage of these properties. Thus, for instance, for plane self-adjoint 2nd order elliptic partial differential equation problems, which after discretization by finite elements or finite differences lead to a positive definite matrix A, the following qualitative figure can be given (cf. Section 7.5).

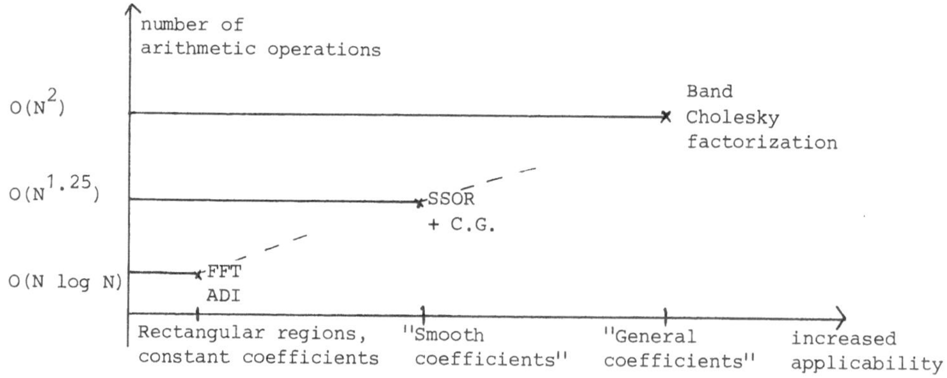

Figure 1.1

We have used the following notations:

N number of unknowns

FFT fast Fourier transform technique

ADI alternating direction implicit method - theory applicable to commutative splittings of A or to separable problems

SSOR symmetric successive overrelaxation method

C.G. conjugate gradient method

Thus in order to prove useful results and to be able to make comparisons between commonly used direct and iterative methods, we will here only deal with *symmetric and positive definite matrices*. Furthermore, in order to *prove* the efficiency of (but not to *use*) the SSOR preconditioning we will need that a further condition is satisfied. This condition is satisfied in many finite element matrix problems (see Section 5.4).

For positive definite matrices A the conjugate gradient method has been used both as a *terminating* (direct) method - theoretically at most N steps are necessary - and as an *iterative* method. We will use the method as an iterative method and as such the number of iterations to reach a given relative accuracy is directly proportional to the square root of the spectral condition number, $\mathcal{H}(A)$, of A, the quotient between the largest and the smallest eigenvalue of A. For special clusterings of the eigenvalues, we will prove that the method may converge faster, however.

The conjugate gradient method is perhaps best presented as an *optimization method* which in fact is not only applicable to linear systems of equations but also to nonlinear systems of equations for which an appropriate corresponding functional to be minimized is given (see Section 2). As such, the method has been used by Hestenes [2], Hestenes and Stiefel [3], Fletcher and Reeves [4], Polak and Ribiére [5], Daniel [6] and by Bartels and Daniel [7] among others.

For linear systems of equations, there are two conjugate gradient algorithms in use :

(*i*) the above mentioned one-step optimization algorithm with updating of search directions

(*ii*) the two-step algorithm

which both have orthogonal (or conjugately orthogonal) residuals.

There is a relationship with the Chebyshev semi-iterative method which will be made clear in Section 3. Actually, a bound on the the rate of convergence of the conjugate gradient method is easily proved using Chebyshev min-max theory, both in the classical case with eigenvalues considered in a single interval of the positive real axis and in the more general case where two or more intervals are considered (see Section 4).

Motivated by the result on the rate of convergence, some so called preconditionings of the matrix A will be mentioned. The term *preconditioning* in connection with the Chebyshev semi-iterative method was used by Evans [8]. The technique was used by D'Yakonov [9] (1961), Habetler and Wachspress [10] (1961), Gunn [11], Dupont et al. [12] and also in connexion with the conjugate gradient method by Axelsson [13] (1972), [14], [15] and [16]. Most of them have used a method similar to the SSOR method as a preconditioning device. Young [17] also used the SSOR method for preconditioning. A numerical comparison between the Chebyshev semi-iterative and the conjugate gradient methods as accelerating devices, based on the (generalized) SSOR method, is made in [13] and [15]. In Bartels and Daniel [7], the discretized Laplacian was used as a preconditioning device (and the Laplacian was "inverted" by FFT-techniques). Later Concus et al. [51] have reviewed some different preconditioning techniques and they use the term "generalized conjugate gradient method". In Section 5 a review of some preconditioning techniques will be given and a simple theory for the SSOR preconditioning will be presented.

Iterative methods for some special matrices will be considered and in Section 7 a review of iterative methods for the classical discretized second-order elliptic problem is given as well as some more recent results for this problem. Finally some numerical results are presented.

2. The classical conjugate gradient algorithm.

Consider the problem of minimizing a functional $f = f(u)$ of N variables, $u \in R^N$. Let $g = g(u) = \text{grad } f(u)$ be the gradient and $H = H(u) = [\frac{\partial^2 f}{\partial u_i \partial u_j}]$ be the Hessian matrix of f. We suppose that H is uniformly positive definite. In order to minimize f, the conjugate gradient method is applicable. In this method the search directions, along which f is successively minimized, are chosen in such a way that they are mutually and conjugately orthogonal (see below).

Before presenting the algorithm, let us mention that such problems are of great importance in physical and engineering sciences where elliptic partial differential problems may be formulated as variational problems; i.e. where the minimal solution of so called *energy functionals*, discretized with finite element approximations, is to be calculated over the corresponding finite dimensional space S_N.

Thus the energy functional usually has such a form that

$$\text{grad } f(u) = A(u)u - b(u),$$

where $A = A(u)$ is the so called "stiffness" matrix given by

$$A_{ij} = \iint_{\Omega} a(u)\ \nabla\phi_i^T(x)\ \nabla\phi_j(x)\ dx\ .$$

$a = a(u) > 0$ is a material coefficient (for instance the diffusion coefficient),

$$u = \sum_{i=1}^{N} u_i\ \phi_i(x), \qquad S_N = \text{SPAN } \{\phi_i(x)\}_{i=1}^{N},$$

ψ_i being the basis functions, and $b = b(\alpha)$ comes from the "source-terms". The equilibrium is thus reached at a point where

$$A(u)u = b(u),$$

this being a necessary condition for extremum. This is in general a nonlinear system of equations. Since the Hessian is supposed to be positive definite, it is also a sufficient condition, and we have a unique solution.

In a linear problem, a is constant and thus A is a constant matrix, independent of the solution vector. Furthermore b is constant. We will for simplicity assume that b is constant in the following but will occasionally make remarks on how to handle problems with a non-constant matrix A.

In a linear problem, the corresponding functional is quadratic,

$$(2.1) \qquad f(u) = \frac{1}{2} u^T A u - b^T u$$

and at the equilibrium point

$$(2.2) \qquad \text{grad } f(u) = Au - b = 0 .$$

This linear system of equations is the well known "equilibrium" equations arising from finite elements applied to linear problems.

For various reasons it may be preferable to consider the problem of minimizing (2.1) instead of considering the solution of (2.2) directly.

One reason for dealing with the functional f is that one can easily add penalty terms, for instance taking care of some boundary constraints. Another reason is that when using the conjugate gradient method in order to minimize the functional, f(u) is *minimized in each iterative step* using the information supplied so far during the iterations (namely the different gradients). This rather vague statement will become more clear in the following.

2.1. The conjugate gradient method as an optimization method.

We will thus now derive the conjugate gradient algorithm from an optimization point of view. However we will consider a somewhat more general functional, enabling us to minimize for instance also the residuals.

Thus let

$$f(u) = \frac{1}{2} u^T A^{\nu} u - (A^{\nu-1} b)^T u , \qquad u \in R^N,$$

where ν is a natural number. Then $g(u) = \text{grad } f(u) = A^{\nu} u - A^{\nu-1} b = A^{\nu-1} r$, where $r = Au - b$ is the residual. We observe that f(u) and g(u) are calculable for all such values of ν. We have

$$f(u) = \frac{1}{2} (u - \hat{u})^T A^{\nu} (u - \hat{u}) - \frac{1}{2} \hat{u}^T A^{\nu} \hat{u}$$

where $\hat{u} = A^{-1} b$ is the solution.

Since the last term is constant, minimizing f(u) is equivalent to minimizing the error functional

$$(2.3) \qquad E(u) = \frac{1}{2} (u - \hat{u})^T A^{\nu} (u - \hat{u}) = \frac{1}{2} r^T A^{\nu-2} r = \frac{1}{2} g(u)^T A^{-\nu} g(u) .$$

For $\nu = 2$ we will thus minimize the Euclidian norm of the residual.

To minimize E (or f) we let $\{d^k\}_{k\geq 0}$ be search directions and $\{\lambda_k\}_{k\geq 0}$ the parameters of exact line search. From

$$u^{k+1} = u^k + \lambda_k d^k , \qquad k = 0,1,\ldots$$

we obtain immediately

$$r^{k+1} = r^k + \lambda_k A\, d^k$$

Since the optimal value of λ_k makes g^{k+1^T} orthogonal to d^k we have

$$0 = g^{k+1^T} d^k = \text{grad } f(u^k + \lambda_k d^k)^T d^k = r^{k+1^T} A^{\nu-1} d^k$$

and hence

$$\lambda_k = - r^{k^T} A^{\nu-1} d^k / d^{k^T} A^{\nu} d^k . \tag{2.4}$$

We now let

$$d^{k+1} = -r^{k+1} + \beta_k d^k , \qquad k = 0,1,\ldots, \qquad d^0 = -r^0$$

i.e., we will in the following conjugate gradient iterative step move along a plane determined by the residual in the last point and the just used search direction. The parameter β_k will be determined later on.

Apparently, since $d^0 = -r^0$, we have

$$d^{k+1} \in \text{SPAN } \{r^0,\ldots,A^{k+1} r^0\},$$

i.e., a linear combination of elements in the Krylov sequence $r^0, Ar^0, \ldots, A^{k+1} r^0$, and likewise

$$r^{k+1} \in \text{SPAN } \{r^0,\ldots,A^{k+1} r^0\}.$$

Thus r^k can be written as a polynomial of degree k in A times r^0 (with constant coefficient = 1), i.e.

$$r^k = (I + P_k(A)) r^0 ,$$

where $P_k(0) = 0$. Thus

$$E(u^k) = \frac{1}{2} r^{k^T} A^{\nu-2} r^k = \frac{1}{2} r^{0^T} (I + P_k(A)) A^{\nu-2} (I + P_k(A)) r^0 . \tag{2.5}$$

Introducing the norm (we observe that A is positive definite)

$$\| u \|_{A^{\nu-2}} = (u^T A^{\nu-2} u)^{1/2} ,$$

we have

$$(E(u^k))^{1/2} = \frac{1}{\sqrt{2}} \| r^0 + P_k(A) r^0 \|_{A^{\nu-2}} .$$

Thus $-P_k(A)r^0$, where $P_k(0) = 0$, can be considered as an approximation of r^0 with error $(E(u^k))^{1/2}$. To minimize this error, it is well known from Hilbert space theory that the error has to be orthogonal (with respect to the corresponding inner product $\langle u,v\rangle = u^T A^{\nu-2} v$) to all functions (vectors) in the linear subspace of approximating functions ; i.e.,

$$\{[I + P_k(A)]r^0\}^T A^{\nu-2} P_l(A) r^0 = 0 , \qquad l \le k$$

or equivalently

$$\{[I + P_k(A)]r^0\}^T A^{\nu-1} [I + P_l(A)] r^0 = 0 , \qquad l \le k-1 .$$

Thus

$$r^{k^T} A^{\nu-1} r^l = 0 , \qquad l \le k-1 ,$$

i.e., the residual vectors will be mutually orthogonal for $\nu = 1$ and in general will be *conjugately orthogonal* with respect to the matrix $A^{\nu-1}$. We will now show that this implies that the search directions will be conjugately orthogonal with respect to the matrix A^ν, i.e. $d^{k^T} A^\nu d^l = 0$, $l \neq k$.

Thus let $l < k$. Then

$$d^{k^T} A^\nu d^l = (A\, d^k)^T A^{\nu-1} d^l = \frac{1}{\lambda_k} (r^{k+1} - r^k)^T A^{\nu-1} d^l$$

$$= \frac{1}{\lambda_k} (r^{k+1} - r^k)^T A^{\nu-1} (-r^l + \beta_{l-1} d^{l-1}) = \frac{\beta_{l-1}}{\lambda_k} (r^{k+1} - r^k)^T A^{\nu-1} d^{l-1} ,$$

where we have used the conjugacy between the residual vectors. By induction,

$$d^{k^T} A^\nu d^l = \frac{1}{\lambda_k} (r^{k+1} - r^k)^T A^{\nu-1} d^l = \frac{\beta_{l-1}}{\lambda_k} (r^{k+1} - r^k)^T A^{\nu-1} d^{l-1} = \ldots =$$

$$= \beta_{l-1}\beta_{l-2} \ldots \beta_0/\lambda_k (r^{k+1} - r^k)^T A^{\nu-1} d^0 ,$$

which is = 0, *since* $d^0 = -r^0$. From this conjugacy, we have in particular

$$(-r^{k+1} + \beta_k d^k)^T A^\nu d^k = 0 ,$$

i.e.,

$$\beta_k = \frac{r^{k+1^T} A^\nu d^k}{d^{k^T} A^\nu d^k} .$$

Since due to exact line searches (which is achieved in a linear problem with the given λ_k-value) we have $d^{k^T} g^{k+1} = r^{k+1^T} A^{\nu-1} d^k = 0$; i.e., with

(2.6) $$d^k = -r^k + \beta_{k-1} d^{k-1} = -r^k + \beta_{k-1}(-r^{k-1} - \dots)$$

we get

$$\beta_k = \frac{r^{k+1^T} A^{\nu-1}(r^{k+1} - r^k)}{d^{k^T} A^{\nu-1}(r^{k+1} - r^k)} = \frac{r^{k+1^T} A^{\nu-1}(r^{k+1} - r^k)}{r^{k^T} A^{\nu-1} r^k} .$$

This formula with $\nu = 1$ (due to Polak, Ribiére [5]) is recommended by Powell [18] for more general (non quadratic) optimization problems. For the above considered quadratic problem it can further be simplified to

$$\beta_k = \frac{r^{k+1^T} A^{\nu-1} r^{k+1}}{r^{k^T} A^{\nu-1} r^k} = \frac{g^{k+1^T} r^{k+1}}{g^{k^T} r^k} .$$

We observe that in the last expression, the Hessian is not needed. Furthermore, from (2.6) we can simplify (2.4),

$$\lambda_k = g^{k^T} r^k / d^{k^T} A^\nu d^k .$$

2.2. The conjugate gradient algorithm.

For an arbitrary initial approximation u^0, the algorithm thus takes the following form:

```
     u:= u^0;
     r:= Au - b;  g:= A^(ν-1) r;  d:= -r;  δ0:= g^T r;
R:   λ:= δ0/d^T A^ν d;
     u:= u + λd;
     r:= Au - b;  g:= A^(ν-1) r;  δ1:= g^T r;
     β:= δ1/δ0;  δ0:= δ1;
     d:= -r + βd;
     IF  r^T r > ε  THEN GOTO  R;
```

For $\nu = 1$ we get the classical conjugate gradient algorithm (cf. [4]), and in variational formulation of elliptic partial differential equations (2.1) then represents the energy in the given system. $\nu = 1$ is thus the appropriate choice. The stopping criterion however ought perhaps be a test on a small enough change in f instead of the test on the residuals. We also observe that $g = r$ and $r^T r = \delta 1$ in that case, so some simplifications are possible in the algorithm (see Section 5).

For $\nu = 2$ we will minimize the Euclidian norm of the residuals, and this may be more natural in other applications than the one just mentioned. Then we also have a stopping criterion in the iteration cycle of a quantity which is to be minimized.

Remark. Instead of calculating the residual according to the definition as done above, it is in a linear problem possible to use the recursion formula

$$r := r + \lambda A d .$$

In the case $\nu = 1$, this will decrease the number of matrix-vector multiplications at each iterative step from 2 to 1 at the expense of having to store one extra vector, namely Ad.

Reid [19] has done a numerical comparison between these two approaches and found only minor differences in the true residual and the one calculated by the recursion formula. The number of vectors to be stored is 3 (u,r,d) compared to 4. (For further results on the computational complexity see [19].)

3. The Chebyshev semi-iterative method.

In this section we will prove that the rate of convergence of the Chebyshev semi-iterative method is determined by the spectral condition number λ_0/λ_1, where λ_1, λ_0 are the smallest and largest of the eigenvalues of A. This will then enable us to give an upper bound on the number of iterations needed in the conjugate gradient method (see Section 4).

3.1. The one-step Chebyshev semi-iterative method.

Let us at first consider the following one-step iterative procedure

$$(3.1) \qquad u^{l+1} = u^l - \tau_{l+1}(Au^l - b), \qquad l = 0,1,\ldots$$

for the solution of $Au = b$. Here $\{\tau_l\}$ is a parameter set, the proper choice of which gives a possible accelerated convergence over the simplest choice $\tau_l = \tau$, $l = 1,2,\ldots$ With $\tau = 2/(\lambda_0 + \lambda_1)$ we then have the smallest spectral radius $\rho(I - \tau A) = (1 - \lambda_1/\lambda_0)/(1 + \lambda_1/\lambda_0)$. The relative error in the Euclidian norm is then decreased to a number at most $\varepsilon > 0$, if $[(1 - \lambda_1/\lambda_0)/(1 + \lambda_1/\lambda_0)]^p \leq \varepsilon$, and this inequality is satisfied if

$$p \geq \frac{1}{2}\frac{\lambda_0}{\lambda_1}\ln\frac{1}{\varepsilon}.$$

The number of necessary iterations are thus in general directly proportional to the spectral condition number of A.

To get an accelerated convergence, we choose a suitable set $\{\tau_l\}$ in (3.1) in order to minimize the corresponding iteration matrix achieved after p iterations. Then the errors satisfy

$$e^p = Q_p(A)e^0$$

where

$$e^p = u^p - u,$$

$$Q_p(\lambda) = \prod_{l=1}^{p}(1 - \tau_l\lambda),$$

and $Q_p(A)$ is the corresponding matrix polynomial. We observe that $Q_p(0) = 1$. We denote by Π_p^1, Π_p^0 the set of polynomials of degree at most p that are = 1 and = 0, respectively, at the origin.

More generally, we would like to minimize the residual $r^p = Au^p - b$, or the error $e^p = A^{-1}r^p$, in the norm

$$\| v \|_{A^{-\nu}} = (v^T A^{-\nu} v)^{1/2} , \qquad \nu \text{ an integer.}$$

Then we have

$$\| r^p \|_{A^{-\nu}} \le \| Q_p(A) \|_{A^{-\nu}} \| r^0 \|_{A^{-\nu}} ,$$

where, as is easily demonstrated,

$$\| Q_p(A) \|_{A^{-\nu}} = \max_i |Q_p(\lambda_i)| ,$$

which we thus want to minimize.

It is well known that the least maximum is achieved by the Chebyshev polynomials, namely

(3.2) $$\min_{Q_p \in \Pi_p^1} \max_{\lambda_1 \le \lambda \le \lambda_0} |Q_p(\lambda)| = \max_{\lambda_1 \le \lambda \le \lambda_0} \frac{|T_p((\lambda_0 + \lambda_1 - 2\lambda)/(\lambda_0 - \lambda_1))|}{T_p((\lambda_0 + \lambda_1)/(\lambda_0 - \lambda_1))}$$

$$= 1/T_p((\lambda_0 + \lambda_1)/(\lambda_0 - \lambda_1)) ,$$

where

$$T_p(z) = \frac{1}{2} [(z + \sqrt{z^2-1})^p + (z - \sqrt{z^2-1})^p] .$$

The optimal choice of the parameters τ_l are thus given by the zeros of T_p, i.e.,

$$\frac{1}{\tau_l} = \frac{\lambda_0 - \lambda_1}{2} \cos \theta_l + \frac{\lambda_0 + \lambda_1}{2} ,$$

where

$$\theta_l = \frac{2 \cdot l - 1}{2p} \pi , \qquad l = 1,2,\ldots,p.$$

In practice, the smallest and largest eigenvalues are not known, so we need lower and upper bounds, a (a > 0) and b, respectively. Then we have to use the parameters

$$\frac{1}{\tau_l} = \frac{b-a}{2} \cos \theta_l + \frac{b+a}{2} .$$

It is an easy matter to find that

$$1/T_p((b + a)/(b - a)) \leq 2\left(\frac{1 - \sqrt{a/b}}{1 + \sqrt{a/b}}\right)^p ,$$

so that if

$$p \geq \frac{1}{2}\sqrt{\frac{b}{a}} \ln \frac{2}{\varepsilon} , \qquad \varepsilon > 0$$

then

$$\min_{Q_p \in \Pi_p^1} \max_{a \leq \lambda \leq b} |Q_p(\lambda)| \leq \varepsilon ,$$

i.e., the relative error in the norm $\|\cdot\|_{A^{-\nu}}$ is at most ε after a cycle of p iterations, as given above.

We observe that in this process, which defines the classical Chebyshev semi-iterative method or Richardson method (see [20]), we have to choose the value of p in advance. Furthermore, it is easily seen that in the parameter set $\{\tau_l\}$ there is a non-empty set for which the matrices $I - \tau_l A$ have spectral radius much larger than 1. This will cause the process to be numerically unstable unless we use some particular permutations of the parameters (see [21]).

Both these disadvantages may be eliminated in the following way.

3.2. The two-step Chebyshev acceleration method.

Thus consider the following two-step formula

$$\text{(3.3)} \qquad \begin{aligned} u^{l+1} &= \alpha_l u^l + (1 - \alpha_l)u^{l-1} - \beta_l r^l , \qquad l = 1,2,\ldots, \\ u^1 &= u^0 - \frac{1}{2}\beta_0 r^0 \end{aligned}$$

We will show how to choose the parameter set $\{\alpha_l, \beta_l\}$ in order that this process apart from rounding errors gives the same result as the one-step Chebyshev process *for every p*. We have with the already introduced notations,

$$e^l = Q_l(A)e^0$$

and observing that the recursion formula (3.3) is valid for all initial vectors, we get

$$Q_{l+1}(A) - \alpha_l Q_l(A) + \beta_l A Q_l(A) + (\alpha_l - 1)Q_{l-1}(A) = 0 , \qquad l = 1,2,\ldots$$

Comparing this with the recursion formula for the Chebyshev polynomials

$$T_{l+1}(z) - 2zT_l(z) + T_{l-1}(z) = 0 , \qquad l = 1,2,...$$

$$T_0(z) = 1, \qquad T_1(z) = z,$$

we see that if

$$\alpha_l = 1 + T_{l-1}(\tilde{b})/T_{l+1}(\tilde{b})$$

$$\beta_l = \frac{4}{b-a} T_l(\tilde{b})/T_{l+1}(\tilde{b}), \qquad \tilde{b} = (b + a)/(b - a),$$

then

$$Q_l(A) = T_l(Z)/T_l(\tilde{b}),$$

$$Z = \frac{1}{b-a} [(b + a)I - 2A],$$

which is the result we wanted (cf. (3.2)). Trivial calculations now give the recursion formulae

$$\alpha_l = \frac{a+b}{2} \beta_l , \qquad \beta_l^{-1} = \frac{a+b}{2} - \left(\frac{b-a}{4} \right)^2 \beta_{l-1} , \qquad l = 1,2,...$$

$$\beta_0 = 4/(a + b) .$$

It is an easy matter to prove that the $\{\beta_l\}$ sequence is monotonically decreasing and converges towards $\tilde{\beta} = 4/(\sqrt{b} + \sqrt{a})^2$, $l \to \infty$ (see e.g. [22]). In [23] it is actually proved that the same asymptotic rate of convergence is achieved with $\beta_l = \tilde{\beta}$, $\alpha_l = \frac{a+b}{2} \tilde{\beta}$, $l = 1,2,...$. Actually, in a problem where $A = I - B$ and the eigenvalues of B belong to the interval $[-\rho,\rho]$, $1 > \rho = \rho(B)$ (the spectral radius of B), we have

$$a = 1 - \rho , \qquad b = 1 + \rho \quad \text{and}$$

$$\alpha = \beta = \tilde{\beta} = 2/(1 + \sqrt{1 - \rho^2}) ,$$

which is recognized as the parameter ω of the optimal SOR method (see [1]).

We will now finally prove that the two-step version of the Chebyshev semi-iterative method is numerically stable.

Thus consider the homogeneous difference equation corresponding to the two-step formula, i.e.,

$$u^{l+1} - \alpha_l u^l + \beta_l A u^l + (\alpha_l - 1)u^{l-1} = 0 \ , \qquad l = 1,2,\dots$$

where u^0 and u^1 are arbitrary vectors. Written as a system of first-order equations we get with

$$W^l = \begin{bmatrix} u^l \\ u^{l-1} \end{bmatrix} ,$$

$$W^{l+1} = \begin{bmatrix} \alpha_l I - \beta_l A & (1 - \alpha_l) I \\ I & 0 \end{bmatrix} W^l$$

The homogeneous difference equation is asymptotically stable if the eigenvalues of this matrix are < 1 in modulus. Denoting the typical pair of eigenvalues of this matrix by $z = z_1(\lambda)$, $z = z_2(\lambda)$, where λ is an eigenvalue of A, we find that z_1 and z_2 must satisfy

$$z^2 - (\alpha_l - \beta_l \lambda) z + \alpha_l - 1 = 0 \ .$$

But it is an easy matter to see that the roots of equation

$$z^2 - \gamma z + \delta = 0$$

are less than 1 in modulus if and only if $|\delta| < 1$ and $|\gamma| < 1 + \delta$

With $\gamma = \alpha_l - \beta_l \lambda$, $\delta = \alpha_l - 1$ the inequalities

$$0 < \alpha_l < 2 \ , \qquad -\alpha_l < \alpha_l - \beta_l \lambda < \alpha_l$$

or

$$0 < \alpha_l < 2 \ , \qquad 0 < \beta_l \lambda < 2\alpha_l \ ,$$

are thus necessary and sufficient for asymptotic stability. These conditions are satisfied since

$$0 < \alpha_l = \frac{b+a}{2} \beta_l < \frac{b+a}{2} \beta_0 = 2 \ , \qquad l \geq 1$$

and, since $0 < a \leq \lambda \leq b$, we also have

$$0 < \beta_l \lambda = \frac{2\lambda}{b+a} \alpha_l \leq 2\alpha_l \ .$$

To summarize, we thus note that in the two-step Chebyshev semi-iterative method the error is of the form

$$\| e^1 \|_{A^{-\nu}} \leq \| Q_1(A) \|_{A^{-\nu}} \| e^0 \|_{A^{-\nu}} ,$$

where $\max |Q_1(\lambda)|$ is minimized over the interval [a,b] in each step (and not only after a fixed number of steps as it was in the one-step semi-iterative formula). Since we deal with a two-step recursion formula it could happen that the "parasite" solution may cause numerical instability. We have proved that this will not happen.

In analogy with (3.3) we may also construct a two-step version of the conjugate gradient method. This will be dealt with in the next chapter in connexion with pre-conditioning of A.

We finally note that at the expense of a little extra work at each iteration, we have in the conjugate gradient method an algorithm which is the fastest of the form considered in minimizing the error ${r^p}^T A^{\nu-2} r^p = {e^p}^T A^{\nu} e^p$ and, perhaps still more important, which *does not demand any estimation of lower and upper bounds of the spectrum of A*.

The conjugate gradient method is limited to inner product (Hilbert) spaces however, whereas the Chebyshev semi-iterative method may be applied in more general Banach spaces.

4. Rate of convergence of the conjugate gradient method.

We will now study the rate of convergence of the conjugate gradient method for linear problems, or equivalently, for quadratic minimization problems where the Hessian $H = A$ is constant. For this purpose we let $\{v_i\}$ be the complete *set* of orthonormal eigenvectors of the symmetric positive definite matrix A; i.e., we have $Av_i = \lambda_i v_i$, where $\{\lambda_i\}$ are the corresponding eigenvalues which we order in a nondecreasing sequence. Denote by S(A) the spectrum of A. Suppose that

$$r^0 = \sum_{i=1}^{N} a_i v_i$$

where $a_i = r^{0^T} v_i$ are the Fourier coefficients of the initial residual (see Section 2). Then from (2.5) we get

$$E(u^k) = \frac{1}{2} r^{k^T} A^{\nu-2} r^k = \frac{1}{2} \min_{p_k \in \Pi_k^0} \sum_{i=1}^{N} a_i^2 [1 + p_k(\lambda_i)]^2 \lambda_i^{\nu-2}$$

$$\leq \frac{1}{2} \min_{p_k \in \Pi_k^0} \max_{\lambda \in S(A)} |1 + p_k(\lambda)|^2 \sum_{i=1}^{N} a_i^2 \lambda_i^{\nu-2} \leq \min_{p_k \in \Pi_k^0} \max_{\lambda_1 \leq \lambda \leq \lambda_0} |1 + p_k(\lambda)|^2 E(u^0)$$

$$\leq E(u^0)/T_k\left(\frac{\lambda_0 + \lambda_1}{\lambda_0 - \lambda_1}\right)^2 .$$

In the last inequality, we have used the result of Section 3 on Chebyshev polynomials. As in the Chebyshev semi-iterative method we thus find that if

$$k = \text{int}\,[\tfrac{1}{2}\sqrt{\frac{\lambda_0}{\lambda_1}}\ \ln \frac{2}{\varepsilon} + 1]$$

the relative error

$$[E(u^k)/E(u^0)]^{1/2} \leq \varepsilon .$$

This upper bound for the number of necessary iterations to reach a relative error of size ε is thus directly proportional to the square root of the spectral condition number

$$\mathcal{H}(A) = \lambda_0/\lambda_1$$

of A, and of course to the number of correct digits.

We observe, however, that the above bound was found by considering the interval $[\lambda_1,\lambda_0]$ as continuous; i.e., by disregarding the discreteness and possible clustering of the eigenvalues for which $a_i \neq 0$. Furthermore, if some of the Fourier coefficients a_i are small an even better estimate could have been found by making use of different sizes of the a_i's and demanding $E(u^k)$ to be less than a prescribed (absolute) small number rather than the relative bound used above. Here we will show that for certain *clusterings* of the eigenvalues we may get better bounds on the number of iterations needed.

4.1. Spectrum contained in two intervals of equal length.

Assume at first that the eigenvalues are distributed in two intervals $[a,b] \cup [c,d]$, $0 < a < b < c < d$. We also assume that $b-a = d-c$. Then

$$p(\lambda) = 1 - \beta\lambda(a + d - \lambda)$$

satisfies $p(a) = p(d)$ and we determine β such that $-p(c) = p(d)$; i.e., $\beta^{-1} = \frac{1}{2}[d(c+a) - c(c-a)]$. Then

$$0 < 1 - p(a) = 1 - p(d) \leq 1 - p(\lambda) \leq 1 - p(b) = 1 - p(c) ,$$

$$\lambda \in S = [a,b] \cup [c,d] ,$$

and $p(\lambda)$ minimizes on S the maximum modulus of all polynomials in Π_2^1. Furthermore, the following polynomial

$$T_k\left(\frac{\tilde{\beta} + \tilde{\alpha} - 2(1 - p(\lambda))}{\tilde{\beta} - \tilde{\alpha}}\right) / T_k\left(\frac{\tilde{\beta} + \tilde{\alpha}}{\tilde{\beta} - \tilde{\alpha}}\right) , \tag{4.1}$$

where $\tilde{\beta} = 1 - p(b)$, $\tilde{\alpha} = 1 - p(a)$, is apparently the best such polynomial in Π_{2k}^1. (The above result is generalized to several intervals in [24].)

If the conjugate gradient method is applied to a linear system of equations with a matrix A with such an eigenvalue distribution, the method needs at most $\text{int}[\sqrt{\frac{\tilde{\beta}}{\tilde{\alpha}}} \ln \frac{2}{\varepsilon} + 1]$ iterations to make $[E(u^k)/E(u^0)]^{1/2} \leq \varepsilon$, where

$$\frac{\tilde{\beta}}{\tilde{\alpha}} = \frac{c}{d}\frac{b}{a} < \frac{b}{a} ,$$

since $1/2 < c/d \leq 1$. This follows since the conjugate gradient method converges in the corresponding norm at least as fast as a convergence acceleration based on any polynomial, including the polynomial (4.1). We call $4\beta/\alpha$ the *effective spectral condition* number of the conjugate gradient method for such a matrix. If $d \gg b$, this number is much smaller than the spectral condition number d/a, valid if we had considered the eigenvalues to belong to just the whole interval [a,d].

If the eigenvalues in the rightmost interval are few enough, we can get an even better upper bound on the necessary number of iterations in the conjugate gradient method. Thus let us now assume that the eigenvalues belong to an interval $[\lambda_1,\lambda_0]$, $\lambda_1 > 0$, and that there are μ additional eigenvalues λ_i', $i = 1,\dots,\mu$ larger than λ_0. Then with

$$1 - q_k(\lambda) = \prod_{i=1}^{\mu} (1 - \frac{\lambda}{\lambda_i'}) \, T_{k-\mu}\left(\frac{\lambda_0 + \lambda_1 - 2\lambda}{\lambda_0 - \lambda_1}\right) / T_{k-\mu}\left(\frac{\lambda_0 + \lambda_1}{\lambda_0 - \lambda_1}\right)$$

we get an upper bound from

$$\max |1 - q_k(\lambda)| = 1/T_{k-\mu}\left(\frac{\lambda_0 + \lambda_1}{\lambda_0 - \lambda_1}\right), \qquad \lambda \in S = [\lambda_1,\lambda_0] \bigcup_{i=1}^{\mu} \lambda_i' .$$

As before,with

$$k = \mathrm{int}\, [\tfrac{1}{2}\sqrt{\frac{\lambda_0}{\lambda_1}} \ln \frac{2}{\varepsilon} + \mu + 1]$$

we have a relative error $< \varepsilon$.

Thus, if μ is at most of the same size as $(\lambda_0/\lambda_1)^{1/2}$, λ_0/λ_1 is the asymptotically effective spectral condition number of the (preconditioned) conjugate gradient method applied to A. We observe that in this latter case we do not need to assume that the eigenvalues λ_i' belong to some interval [c,d] such that $d - c \geq b - a$ and, more important, such that $d + a \leq 2c$.

4.2. Spectrum contained in two arbitrary positive intervals.

We now consider two arbitrary positive intervals. Thus let the eigenvalues belong to two intervals, where the condition number d/c of the interval to the right, [c,d], may be large. (In Section 4.1, $d/c \leq 2$.) Further, we assume that the interval to the left, [a,b], contains μ eigenvalues. To get an upper bound on the number of iterations in the conjugate gradient method we let

$$1 - q(x) = r_m(x) \; T_{k-m} \left(\frac{d + c - 2x}{d - c}\right) / \; T_{k-m} \left(\frac{d + c}{d - c}\right)$$

where $r_m \in \Pi_m^1$. Then $q \in \Pi_m^0$ and

$$\max_{x\in[a,b]} |1 - q(x)| \leq \max_{x\in[a,b]} |r_m(x)| ,$$

$$\max_{x\in[c,d]} |1 - q(x)| \leq \max_{x\in[c,d]} |r_m(x)| \; / \; T_{k-m} \left(\frac{d + c}{d - c}\right) .$$

Let the eigenvalues

$$\lambda_i \in [c,d] \bigcup_{i=1}^{\mu} \lambda_i^! , \qquad \lambda_i^! \in [a,b] .$$

If μ is small enough then we choose

$$(i) \qquad r_m(x) = \tilde{r}_m(x) = \prod_{i=1}^{\mu} (1 - x/\lambda_i^!) , \qquad m = \mu.$$

Thus $r_m(x) = 0, \ x \in \bigcup_{i=1}^{\mu} \lambda_i^!$. Otherwise we choose

$$(ii) \qquad r_m(x) = \approx{r}_m(x) = T_m \left(\frac{b + a - 2x}{b - a}\right) / T_m \left(\frac{b + a}{b - a}\right) .$$

(This choice was suggested in [25].)

We have

$$\max_{[a,b]} |\tilde{r}_m(x)| \leq \left(\frac{b}{a} - 1\right)^m = \eta_1$$

and

$$\max_{[a,b]} |\approx{r}_m(x)| \leq 2\left(\frac{\sqrt{b} - \sqrt{a}}{\sqrt{b} + \sqrt{a}}\right)^m = \eta_2 .$$

In the first case $m = \mu$ and in the latter case we choose

$$m = \text{int} \left[\frac{1}{2}\sqrt{\frac{b}{a}} \ln \frac{2}{\varepsilon} + 1\right], \text{ so that}$$

$$\max_{[a,b]} |r_m(x)| \leq \varepsilon.$$

Furthermore, since the Chebyshev polynomial $T_m\left(\frac{2x - b - a}{b - a}\right)$ grows fastest outside the interval $[a,b]$ of all polynomials of degree m whose maximum absolute value in $[a,b]$ is 1 (see for instance [26]), we have

$$\max_{d \geq x \geq b} |r_m(x)| \leq \eta \, T_m\left(\frac{2d - b - a}{b - a}\right) ,$$

$\eta = \eta_1, \eta_2$ respectively. Thus in both cases (i) and (ii) we get from $T_m(\zeta) \leq (2\zeta)^m$, $\zeta \geq 1$ that

$$\max_{x \in [c,d]} |1 - q(x)| \leq \varepsilon$$

if

$$\eta\left(\frac{4d}{b - a}\right)^m / T_{k-m}\left(\frac{d + c}{d - c}\right) \leq \varepsilon ,$$

i.e., if

$$T_{k-m}\left(\frac{d+c}{d-c}\right) \geq \frac{2}{\varepsilon}\left(\frac{4d}{e}\right)^m$$

where e = a and e = b for case (*i*) and (*ii*), respectively.
As before, we find that this inequality is satisfied if

$$k - m \geq \frac{1}{2}\sqrt{\frac{d}{c}}\left(\ln\frac{2}{\varepsilon} + m \ln\frac{4d}{e}\right)$$

or if

$$k = \text{int}\left[\frac{1}{2}\sqrt{\frac{d}{c}} \ln\frac{2}{\varepsilon} + \left(1 + \frac{1}{2}\sqrt{\frac{d}{c}} \ln\frac{4d}{e}\right)m\right] \tag{4.2}$$

where, since we may choose the analysis in case (*i*) or (*ii*), whichever is most convenient,

$$1 \leq m = \min\left(\frac{1}{2}\sqrt{\frac{b}{a}} \ln\frac{2}{\varepsilon}, \mu\right) .$$

If $m = \mu$ (case (*i*)) then e = a, else e = b (case (*ii*)).

If $b/a \gg 1$ but μ is large enough so that $m < \mu$ then

$$k \sim \frac{1}{4}\sqrt{\frac{b}{a}}\sqrt{\frac{d}{c}} \ln\frac{4d}{b} \ln\frac{2}{\varepsilon} .$$

If $\frac{1}{2}\sqrt{\frac{b}{c}} \ln\frac{4d}{b} < 1$ we have then got an effective spectral condition number which is smaller than the condition number d/a corresponding to the whole interval [a,d]. In this, as well as in the first case, it is obvious how to extend the results to several intervals.

5. The preconditioned conjugate gradient algorithm.

Let us now introduce a square nonsingular matrix C, called the preconditioning matrix, which is assumed to be simple in the respect that systems of linear equations with matrix C should be easily solvable; i.e., need a small number of operations and make a small demand on the size of memory. Thus, for instance, C can be a product of sparse triangular matrices. We will here consider the preconditioning in the (one-step) conjugate gradient method and in the two-step algorithm and also review some preconditioning techniques which have been used. From Section 4 dealing with the rate of convergence, we are guided in the choice of C in the respect that we should try to decrease the resulting spectral condition number (namely of $C^{-1}A$ as we will see).

5.1. The preconditioned algorithm.

For simplicity, in the following we consider only the functional in (2.3) with $\nu = 1$. Let $C = EE^T$ be positive definite and consider the error functional (see (2.3))

$$E(u^k) = \frac{1}{2} g^{k^T} A^{-1} g^k = \frac{1}{2} (E^{-1}g^k)^T (E^{-1}AE^{-T})^{-1} E^{-1} g^k = \frac{1}{2} \tilde{g}^{k^T} \tilde{A}^{-1} \tilde{g}^k ,$$

where

$$\tilde{g}^k = E^{-1} g^k , \qquad \tilde{A} = E^{-1} A E^{-T} .$$

Further, let

$$\tilde{u} = E^T u , \qquad \tilde{b} = E^{-1} b .$$

The conjugate gradient algorithm in Section 2 for the transformed system $\tilde{A}\tilde{u} = \tilde{b}$ is

$\tilde{u} := \tilde{u}^0$;

$\tilde{g} := \tilde{A}\tilde{u} - \tilde{b}$;

$\tilde{d} := -\tilde{g}$; $\delta 0 := \tilde{g}^T \tilde{g}$;

R: $\lambda := \delta 0 / \tilde{d}^T \tilde{A} \tilde{d}$;

$\tilde{u} := \tilde{u} + \lambda \tilde{d}$;

$\tilde{g} := \tilde{A}\tilde{u} - \tilde{b}$; $\delta 1 := \tilde{g}^T \tilde{g}$;

$\beta := \delta 1 / \delta 0$; $\delta 0 := \delta 1$;

$\tilde{d} := -\tilde{g} + \beta \tilde{d}$;

<u>IF</u> $\delta 1 > \varepsilon$ <u>THEN GOTO</u> R;

Going back to the untransformed quantities we get (compare Remark in Section 2)

$u := u^0;\quad g := Au - b;$

$C\gamma := g;$

$e := -\gamma;\quad \delta 0 := g^T\gamma;$

R: $\lambda := \delta 0/e^T Ae;$

$u := u + \lambda e;$

$g := g + \lambda Ae;$

$C\gamma := g;\quad \delta 1 := g^T\gamma;$

$\beta := \delta 1/\delta 0;\quad \delta 0 := \delta 1;$

$e := -\gamma + \beta e;$

<u>IF</u> $\delta 1 > \varepsilon$ <u>THEN</u> <u>GOTO</u> R;

which is the *preconditioned conjugate gradient algorithm* expressed in untransformed vectors. In this version we do not need to make a final back-transformation and in each stage of the iterative process we work with approximations to the solution vector.

In comparison with the unpreconditioned algorithm in Section 2, we need only storage for one additional vector and at each iterative step we have in addition to solve a system of linear equations of the form

(5.1) $\quad C\gamma = g\ .$

If C is an appropriate matrix this is a fast procedure however.

Apparently C is in some way an approximation of A so that the spectral condition number of $C^{-1}A$ is small. (We observe that $C^{-1}A$ is similar to $E^{-1}A\,E^{-T}$ and thus has the same eigenvalues.) At this point one may ask why we do not try to find an appropriate approximation to the inverse A^{-1} itself, thus avoiding the solution of the system of equations (5.1). However A^{-1} is usually a dense matrix and the approximation must then also be a fairly dense matrix. This means that even multiplications with such a matrix may be at least as time-consuming as the solution of (5.1) and at the same time we have lost in sparsity, the main goal with the iterative scheme being to preserve sparsity.

5.2. A two-step preconditioned conjugate gradient algorithm.

Let us now consider the two-step iteration formula,

(5.2) $$C[u^{l+1} - \alpha_l u^l + (\alpha_l - 1)u^{l-1}] = -\beta_l r^l , \qquad l = 0,1,\ldots, \qquad \alpha_0 = 1 .$$

Apparently

$$r^{l+1} = \alpha_l r^l + (1 - \alpha_l) r^{l-1} - \beta_l A\, C^{-1} r^l ,$$

where

$$r^l = A\, u^l - b = (I + p_l(A\, C^{-1})) r^0, \qquad p_l \in \pi_l^0 .$$

Again, in order to study this algorithm we consider the transformed quantities

$$\tilde{r}^l = E^{-1} r^l , \qquad \tilde{A} = E^{-1} A\, E^{-T}$$

and note that $\tilde{r}^k = (I + p_k(\tilde{A}))\tilde{r}^0$. We again assume that $C = EE^T$ is positive definite. It follows then from Section 2 that

$$\tilde{r}^{k^T} \tilde{A}^{-1} \tilde{r}^k = r^k A^{-1} r^k$$

is minimized for all $\tilde{r}^k$ of the above form if

$$\tilde{r}^{l^T} \tilde{r}^j = 0 , \qquad j \leq l-1$$

i.e.,if

$$r^{l^T} C^{-1} r^j = 0 , \qquad j \leq l-1 , \qquad l = 1,2,\ldots$$

It is easily seen that it is possible to determine the coefficients $\{\alpha_l, \beta_l\}$ in the three-term recursion formula (5.2) such that this orthogonality (conjugacy) condition is satisfied. We get

$$r^{(l+1)^T} C^{-1} r^l = 0 \implies \alpha_l = \beta_l \mu_l ,$$

$$\mu_l = (C^{-1} r^l)^T A (C^{-1} r^l)/\delta_l$$

$$\delta_j = r^{j^T} C^{-1} r^j ,$$

and

$$r^{(l+1)^T} C^{-1} r^{l-1} = 0 \implies \beta_l^{-1} = \mu_l - \frac{\delta_l}{\delta_{l-1}} \beta_{l-1}^{-1} .$$

Furthermore, it is easily seen that

$$r^{(l+1)^T} C^{-1} r^j = 0 , \qquad j = l-2, l-3, \ldots, 0$$

with this choice of $\{\alpha_l, \beta_l\}$. For details see [13],[15] and [16]. Compare also [27].

Now μ_l is bounded by the smallest and largest of the eigenvalues of $C^{-1}A$. We will show that $\alpha_l > 1$, $\beta_l > 0$ if $r^l \neq 0$. At first we observe that the following modification of the recursion formula, where we work with differences, is advantageous in order to decrease the influence of rounding errors:

$$C\,[\Delta u^{l+1} - (\alpha_l - 1)\Delta u^l\,] = -\beta_l r^l\,, \qquad u^{l+1} = u^l + \Delta u^l$$

$$C\,[\Delta r^{l+1} - (\alpha_l - 1)\Delta r^l\,] = -\beta_l A r^l\,, \qquad r^{l+1} = r^l + \Delta r^{l+1}\,, \qquad l = 1,2,\ldots \tag{5.3}$$

For notational convenience, we let $C = I$, the identity matrix, in the following.

<u>Lemma:</u> $\quad r^{l^T} A^{-1} \Delta r^{(l-i)} = 0\,, \qquad i = 0,1,\ldots,l\,.$

<u>Proof:</u> $\quad r^{l^T} A^{-1} \Delta r^{(l-i)} = (\alpha_{l-i-1} - 1) r^{l^T} A^{-1} \Delta r^{(l-i-1)} - \beta_{l-i-1} r^{l^T} r^{(l-i-1)}$

$$= (\alpha_{l-i-1} - 1) r^{l^T} A^{-1} \Delta r^{(l-i-1)} = \ldots = \prod_{j=i+1}^{l-1} (\alpha_{l-j} - 1) r^{l^T} A^{-1} \Delta r^1$$

$$= -\beta_0 \Pi(\alpha_{l-j} - 1) r^{l^T} r^0 = 0\,.$$

We now multiply (5.3) in turn by r^l and $A^{-1}(r^l - r^{l-1})$. Then

$$\begin{bmatrix} r^{l^T} A\, r^l & -r^{l^T} r^l \\ -r^{l^T} r^l & \Delta r^{l^T} A^{-1} \Delta r^l \end{bmatrix} \begin{bmatrix} \beta_l \\ \tilde{\alpha}_l \end{bmatrix} = \begin{bmatrix} r^{l^T} r^l \\ 0 \end{bmatrix}$$

where $\tilde{\alpha}_l = \alpha_l - 1$. Thus

$$\begin{bmatrix} \beta_l \\ \tilde{\alpha}_l \end{bmatrix} = \frac{1}{\det} \begin{bmatrix} \Delta r^{l^T} A^{-1} \Delta r^l & r^{l^T} r^l \\ r^{l^T} r^l & r^{l^T} A\, r^l \end{bmatrix} \begin{bmatrix} r^{l^T} r^l \\ 0 \end{bmatrix}$$

where

$$\det = r^{l^T} A\, r^l\, \Delta r^{l^T} A^{-1} \Delta r^l - (r^{l^T} r^l)^2 = r^{l^T} A\, r^l\, \Delta r^{l^T} A^{-1} \Delta r^l - (\Delta r^{l^T} r^l)^2 \geq 0\,,$$

by the Cauchy-Schwarz inequality, using that A is positive definite. In fact $\det > 0$, since if not, $A^{1/2} r^l$ and $A^{-1/2} \Delta r^l$ are linearly dependent, which leads to $r^{l^T} A^{-1} r^l = 0$, i.e. $r^l = 0$. Thus we have $\beta_l > 0$ and $\alpha_l - 1 = \tilde{\alpha}_l > 0$.

Since this conjugate gradient method is based on a three-term recursion formula, the second "parasite" solution may cause numerical instability (cf. [28]). However, although $\beta_1 > 0$ and $\alpha_1 > 1$ are not sufficient conditions for stability (cf. Section 3.2; in fact it is an easy matter to construct examples for which at least some α_1 is > 2) we have some evidence that instability is a rare phenomenon. In practice only extremely large condition numbers seem to make this instability noticeable. Also note that the preconditioning decreases this condition number.

5.3. Some preconditioning techniques.

One of the simplest preconditioning techniques is based on the SSOR method (symmetric successive overrelaxation method; see e.g. [17]). In this

$$C = (D + \omega L)D^{-1}(D + \omega L^T)$$

where $A = D + L + L^T$ and D is the diagonal or a block-diagonal part of A. As a special case we have diagonal scaling (cf. [29]).

A simple theory for the SSOR preconditioning will be given in the next section. The advantage with this preconditioning is that all elements of the factors in C are given from those in A and thus that the sparsity structure in A is preserved.

Since

$$\frac{1}{\omega} C = A + (\frac{1}{\omega} - 1)D + \omega L D^{-1} L^T$$

the method can also be seen as a technique for approximate factorization of A. The rate of convergence with respect to the parameter ω is often quite insensitive to the choice of this parameter (see [15]). Such symmetric factorization methods have been considered in connexion with difference methods for 2nd order partial differential equations in Dupont et al [12], Axelsson [13], and in a more general way in Meijerink and van der Vorst [30] (but without a parameter ω).

The idea in the latter paper is just to include more elements in the approximate LU-decomposition of A (unless these elements are small enough) so that a more accurate approximate factorization of A is achieved. It is still possible to introduce an "SSOR-parameter" in order to even further decrease the condition number. When accelerated by the conjugate gradient method, the dependence on the parameter is minor however (cf. [13]).

In this respect it should however be remarked that it is not a good approximate inverse we are looking for, but a matrix which decreases the spectral condition number. Thus, for instance, it has been found that the Laplacian often works well as a preconditioning of more general discretized elliptic operators: see Gunn [11], Concus and Golub

[31] and Bartels and Daniel [7].

In general, the properties a good preconditioning matrix must satisfy are

- (*i*) The spectral condition number of $C^{-1}A$ should be much smaller than that of A
- (*ii*) the calculation of C (preconditioning work) must not need too much effort
- (*iii*) C should be (almost) as sparse as A
- (*iv*) the computational complexity of the linear equations with coefficient matrix C should be small in comparison with that of A.

As an extreme case, the two-step preconditioned conjugate gradient method includes a method which is usually regarded as a direct method. Namely, if A is considered to be a full bandmatrix, a common choice of C is $C = U^tU$, the Cholesky decomposition of A, where U is upper triangular. Then, apart from rounding errors, $\mathcal{H}(C^{-1}A) = 1$ and we get $\alpha_1 = \beta_1 = 1$ and $x^0 = 0$. This apparently gives us the well known Cholesky method followed by one or several steps of iterative refinement, since in practice the factorization of A is only approximate, due to rounding errors. Apparently, however, in the Cholesky choice of the matrix C, C (or the factors of C) is not sparse (see (*iii*) above) and the calculation of it needs a computational effort which dominates by an order of magnitude the computational complexity involved in the conjugate gradient method (see table 1 in Section 7).

At the other extreme, C = I gives a pure conjugate gradient method (with no preconditioning).

In between these two extremes there is an infinity of other choices of C, satisfying the conditions (*i*) - (*iv*) to different extents and leading to different approximate factorization methods of which some examples have been given above.

5.4. The SSOR preconditioning matrix.

We will first consider a slightly more general preconditioning matrix, particular cases of which are the alternating direction (ADI) and successive overrelaxation (SOR) methods (see [1]). Thus let

$$A = B + H + V$$

be a splitting of A, where B is nonsingular. If the matrices B + H and B + V are easily invertible from a practical point of view, like the product of tridiagonal or triangular matrices, then the matrix

$$C = (B + H)B^{-1}(B + V)$$

is usable for the conditioning of A. To solve the linear system of equations of the form $Cy = g$, we solve successively

$$\begin{cases} (B + H)\, y^{1/2} = g \\ (B + V)\, y^{1} = By^{1/2} \end{cases}$$

Suppose now that the Hermitian transposes $V^* = H$, $B^* = B$ and B is positive definite. Then $C^* = C$, C is positive definite and $C^{-1}A$ is similar to $C^{-1/2} A\, C^{-1/2}$ which is Hermitian. Thus the eigenvalues of $C^{-1}A$ are real. Furthermore, the eigenvalues of $C^{-1}A$ are bounded below and above by lower and upper bounds, respectively, of the quotient $(Ax,x)/(Cx,x)$ of the quadratic forms of A and C (i.e. $(Ax,x) = x^TAx$ etc). This quotient is obviously positive and an upper bound, 1, is immediately found, since

$$C = B + H + V + HB^{-1}V = A + V^* B^{-1}V$$

and thus

$$(Cx,x) \geq (Ax,x) .$$

Consider now the particular choice, the SSOR-method (symmetric successive overrelaxation method, see e.g. [17]) where $A = D + L + U$, D is (block-) diagonal and positive definite, $L = U^*$ is lower triangular and $B = [(2/\omega) - 1]D$, $0 < \omega < 2$. Then with

$$H = (1 - \frac{1}{\omega})D + L , \qquad V = (1 - \frac{1}{\omega})D + U ,$$

we have

$$B + H = \frac{1}{\omega} D + L , \qquad B + V = \frac{1}{\omega} D + U .$$

$$C = \frac{1}{2-\omega} \left(\frac{1}{\omega} D + L\right)\left(\frac{1}{\omega} D\right)^{-1}\left(\frac{1}{\omega} D + U\right) .$$

Here the factor $1/(2-\omega)$ is a matter of convenience in the theoretical calculations only. C depends also on the ordering of the equations. The natural ordering, taking the nodes in some row-wise ordering,works well however. (See remarks made at the end of this section.) One advantage with the SSOR preconditioning is apparently that we do not need to calculate any new matrix coefficients.

We now examine how to choose the parameter ω in some optimal way. After some simple algebraic calculations we get

$$(2 - \omega)C = A + \frac{1}{4\omega} (2 - \omega)^2 D + \omega(LD^{-1}L^* - \frac{1}{4} D) .$$

Suppose now that there exist scalars μ, δ such that (*i*) $\mu = \max_{x \in C^N} \{(Dx,x)/(Ax,x)\}$,

(*ii*) $\delta \geq \min \{0, \max_{x \in C^N} [(LD^{-1}L^* x,x) - \frac{1}{4} (Dx,x)]/(Ax,x)\}$.

Then we have for all N-dimensional vectors x,

$$0 < \frac{2 - \omega}{1 + \mu(2 - \omega)^2/4\omega + \omega\delta} \leq \frac{(Ax,x)}{(Cx,x)} \leq 1 ,$$

i.e., the spectral condition number of $C^{-1}A$ is bounded above by the inverse of the lower bound.

With

$\omega = 2/(1 + \zeta/\sqrt{\mu})$, $\zeta > 0$ a parameter, we get

$$\mathcal{H}(\zeta) = \mathcal{H}(C^{-1}A) \leq \left(\frac{1/2 + \delta}{\zeta} + \frac{\zeta}{4} \right) \sqrt{\mu} + \frac{1}{2} ,$$

which upper bound is minimized for

(5.4) $\zeta = \zeta^* = 2\sqrt{\frac{1}{2} + \delta}$

and then

$$\mathcal{H}(\zeta^*) \leq \sqrt{(\frac{1}{2} + \delta)\mu} + \frac{1}{2} .$$

It is however clear that $\mathcal{H}(\zeta)$ is fairly insensitive to the choice of $\zeta > 0$ (and to the estimate of the number μ). The actual number of necessary iterations to reach a demanded accuracy after application of the conjugate gradient acceleration process is actually even more insensitive to this choice (see [13] and [15]).

We observe that μ is equal to the reciprocal of the smallest eigenvalue of the diagonally scaled matrix $\tilde{A} = D^{-1/2}A\,D^{-1/2}$ and in practice of the same order as the spectral condition number of A. Thus, if the number δ is small enough, we have gained in a smaller spectral condition number by the SSOR preconditioning. We will now show that for a special class of matrices, we have $\delta = 0$. Then the resulting spectral condition number is $O(\sqrt{\mu})$.

Thus, suppose $\tilde{A} = I + \tilde{L} + \tilde{L}^*$ is diagonally dominant with rowsums

(5.5) $\|\tilde{L}\|_\infty \leq \frac{1}{2}$, $\|\tilde{L}^*\|_\infty \leq \frac{1}{2}$.

Then

$$(LD^{-1}L^*x,x) - \frac{1}{4}(Dx,x) \leq (\rho(\tilde{L}\tilde{L}^*) - \frac{1}{4})||\tilde{x}||^2 \leq ||\tilde{L}||_\infty||\tilde{L}^*||_\infty - \frac{1}{4})||\tilde{x}||^2 \leq 0,$$

where $\tilde{x} = D^{1/2}x$, i.e., $\delta = 0$.

It is important at this stage to note that the number δ is dependent on the *ordering* of the equations. Even if (5.5) is satisfied for a particular ordering it may be violated for another ordering. An example of this is given in Chapter 7. The ordering strategy should be to try to have the predecessors and postdecessors to contribute equally in the rowsum.

Finally we remark that with the corresponding theory for the SOR method (where we do not assume consistently ordered systems of equations) the same (optimal) parameter

$$\omega = \tilde{\omega}_{opt} = 2/[1 + 2\sqrt{(\frac{1}{2} + \delta)/\mu}\,]$$

as in (5.4), results in a spectral radius

$$\rho(\mathcal{L}_\omega) \leq 1 - 1/2[\sqrt{(\frac{1}{2} + \delta)/\mu} + \frac{1}{2}\,].$$

For details, see Andersson [25]. This bound is only half as favourable as the corresponding spectral radius in the SSOR method. The straightforward application of the latter, however, requires double the work per iteration. We mention that the SOR method (with optimal parameter ω) cannot be combined with an accelerating method (see Section 6.2) and the interest in the SSOR method lies in the fact that this latter method can be accelerated.

It is of interest to compare this parameter $\tilde{\omega}_{opt}$ obtained for symmetric positive definite matrices with that of the classical SOR method obtained for matrices with property A (see [1], [17]). Then

$$\omega_{opt} = 2/[1 + \sqrt{1 - \rho(\tilde{B})^2}\,]$$

when $\rho(\tilde{B})$ is the spectral radius of the Gauss-Jacobi iteration matrix $\tilde{B} = I - \tilde{A}$. Thus

$$\mu = 1/[1 - \rho(\tilde{B})]$$

and

$$\omega_{opt} = 2/[1 + \sqrt{\frac{2}{\mu}(1 - \frac{1}{2\mu})}\,] > \tilde{\omega}_{opt}\ .$$

If $\delta = 0$, the two parameters are very close.

6. Iterative methods for some special classes of matrices.

In this section iterative methods for some special systems of linear equations will be presented. The methods are combinations of iterative methods; for instance, coupled inner-outer iterations.

6.1. A special system of equations.

Let Q_n be a polynomial of degree n with real coefficients such that $Q_n(\lambda) \neq 0$, $\mathrm{Re}\,(\lambda) > 0$. We would like to solve

$$Q_n(A)x = b \ ,$$

where A is a positive definite operator in a Hilbert space, in particular a positive definite matrix, without calculating any powers of A. Suppose then that Q_n is factorized (according to the fundamental lemma of algebra) in factors of at most second degree, with real coefficients. For each of the quadratic factors we thus get a system of equations of the form

$$(6.1) \qquad Q_2(A)y = b' \ ,$$

$$Q_2(A) = I + \alpha A + \beta A^2 \ , \qquad 4\beta > \alpha^2 \ ,$$

the solution of which we now consider.

Thus let

$$C = (I + \gamma A)^2 = I + 2\gamma A + \gamma^2 A^2 \ , \qquad \gamma = \sqrt{\beta} \ ,$$

be a preconditioning matrix. Then

$$(Q_2(A)v,v)/(Cv,v) = I - (2\gamma - \alpha)(Av,v) \,/\, (Cv,v) \ .$$

We have

$$\frac{(2\gamma - \alpha)(Av,v)}{(Cv,v)} = (1 - \frac{\alpha}{2\gamma}) \frac{(Av,v) + (Av,v)}{\frac{1}{\gamma}||v||^2 + \gamma\,||Av||^2 + 2(Av,v)}$$

$$\leq (1 - \frac{\alpha}{2\gamma}) \frac{\frac{1}{2}[\frac{1}{\gamma}||v||^2 + \gamma\,||Av||^2] + (Av,v)}{\frac{1}{\gamma}||v||^2 + \gamma\,||Av||^2 + 2(Av,v)} = \frac{1}{2}(1 - \frac{\alpha}{2\gamma})$$

Thus

$$\frac{1}{2}(1 + \frac{\alpha}{2\gamma}) < \frac{(Q_2(A)v,v)}{(Cv,v)} < 1$$

i.e., the spectral condition number satisfies

$$(6.2) \qquad \mathcal{H}(C^{-1}Q_2(A)) < \frac{2}{1 + \frac{\alpha}{2\sqrt{\beta}}}$$

which upper bound is *independent of A*. The linear, real factors $(I + \gamma A)$ can be dealt with using some of the previous iterative techniques (inner iterations) or by a direct method. The number of outer iterations

$$(6.3) \qquad C(y^{l+1} - y^l) = -\tau_1(Q_2(A)y^l - b') , \qquad l = 0,1,\ldots$$

are essentially independent of the spectral condition number of A. (6.3) is the preconditioned form of the one-step Chebyshev semi-iterative method applied to the system (6.1).

An application of this technique is in connection with rational (e.g. Padé) approximations of exp(A) for the numerical solution of the evolution equation $u' = Au + f(t)$.

Another application is of course when we would like to solve systems of equations

$$(A + i\omega B)v = f ,$$

where A is positive definite and B is symmetric, working only with real algebra. Let v_1, v_2 and f_1, f_2 be the real and imaginary parts of v and f, respectively. Then by trivial computations

$$[A + \omega^2 B A^{-1} B]v_1 = q ,$$

$$q = \omega B A^{-1} f_2 + f_1 .$$

In order to solve this system of equations by iteration we precondition it by

$$C = (A + \omega B)A^{-1}(A + \omega B) .$$

Letting $\tilde{A} = A^{-1/2} B A^{-1/2}$, $\tilde{v}_1 = A^{1/2} v_1$, $\tilde{q} = A^{-1/2} q$ and $\tilde{C} = A^{-1/2} C A^{-1/2}$ we have

$$[I + \omega^2 \tilde{A}^2]\tilde{v}_1 = \tilde{q} ,$$

$$\tilde{C} = (I + \omega\tilde{A})^2 .$$

According to the upper bound (6.2), the process corresponding to (6.3) will converge at least with a rate of convergence determined by $\mathcal{H} < 2$ or by $\sqrt{\mathcal{H}} < \sqrt{2}$ if an accelerating formula is used.

6.2. The Chebyshev semi-iterative method for matrices with complex eigenvalues.

At this point we may mention that the Chebyshev method may also be applied to matrices with complex eigenvalues.

Thus assume that the eigenvalues of A are situated in an ellipse in the right half plane,

$$E = \{\zeta = 1 - \alpha \cos\theta + i\beta \sin\theta ; \quad 0 \leq \theta \leq 2\pi\} ,$$

where $0 \leq \beta < \alpha < 1$. (The number 1 is just a factor of normalization.) By the transformation

$$z = (1 - \zeta)/\sqrt{\alpha^2 - \beta^2}$$

we obtain a new ellipse

$$E_\rho = \{z = \frac{1}{2} [\rho e^{i\theta} + \rho^{-1} e^{-i\theta}] ; \quad 0 \leq \theta \leq 2\pi\} ,$$

where the sum of the semi-axes is

$$\rho = \frac{\alpha + \beta}{\sqrt{\alpha^2 - \beta^2}} = \sqrt{\frac{\alpha + \beta}{\alpha - \beta}} \geq 1 .$$

From approximation theory (see also e.g. [22]) we have

$$\mathcal{H}_p = \min_{Q_p \in \pi_p} \max_{\zeta \in E} \left| Q_p\left(\frac{1 - \zeta}{\sqrt{\alpha^2 - \beta^2}}\right) / Q_p\left(\frac{1}{\sqrt{\alpha^2 - \beta^2}}\right)\right|$$

$$= \max_{\zeta \in E} \left| T_p\left(\frac{1 - \zeta}{\sqrt{\alpha^2 - \beta^2}}\right) / T_p\left(\frac{1}{\sqrt{\alpha^2 - \beta^2}}\right)\right| ,$$

and T_p is the unique such approximation, apart from a constant factor. Thus

$$|T_p(z)/T_p(1/\sqrt{\alpha^2 - \beta^2})|$$

has the smallest maximum value of all polynomials of degree p over the set E.

Let

$$a = 1 - \sqrt{\alpha^2 - \beta^2} , \qquad b = 1 + \sqrt{\alpha^2 - \beta^2} .$$

Then

$$\frac{1}{b-a} (b + a - 2\zeta) = z$$

and

$$\frac{b+a}{b-a} = 1/\sqrt{\alpha^2 - \beta^2} .$$

Thus the sequence of Chebyshev polynomials in (3.1) (or in (3.3)), defined by

$$\tau_l^{-1} = \frac{b-a}{2}\cos(\theta_l) + \frac{b+a}{2}, \qquad \theta_l = \frac{2l-1}{2p}\pi, \qquad l = 1,\ldots,p$$

(or by (3.3)), will converge fastest of all such polynomial sequences over E.

Since, as is easily proved, using the recurrence relation for Chebyshev polynomials,

$$T_p(z) = \frac{1}{2}(\rho^p e^{ip\theta} + \rho^{-p} e^{-ip\theta}) ,$$

we have

$$\max_{\zeta\in E} |T_p(z)| = \frac{1}{2}(\rho^p + \rho^{-p}) .$$

From the above it follows that the asymptotic average convergence factor is

$$\lim_{p\to\infty} \mathcal{H}_p^{1/p} = \frac{\rho}{\frac{1}{\sqrt{\alpha^2-\beta^2}} + \sqrt{\frac{1}{\alpha^2-\beta^2} - 1}} = \frac{\alpha+\beta}{1+\sqrt{1-(\alpha^2-\beta^2)}}$$

From the above it follows that if $\beta \to \alpha < 1$; i.e., if the ellipse approaches a circle, the best approximation is simply z^p. That is, if the domain containing the eigenvalues of A is assumed to be circular, no Chebyshev acceleration scheme can be devised which will improve the convergence of the given iteration process.

6.3. A coupled Chebyshev semi-iterative method and conjugate gradient method.

It has to be observed that the computational complexity of the conjugate gradient method is slightly larger than that for the Chebyshev method, in particular for a (stable) one-step Chebyshev method. Thus at least two more inner products, the calculation of at least one extra vector (involving multiplication by a constant and addition of two vectors) and storage of some extra vectors are needed in the former method. The relative importance of this depends on the number of non-zero elements in A; the more such elements there are per row, the less important is this extra cost in the conjugate gradient method.

The advantage of the conjugate gradient method, besides being in general faster than the Chebyshev method, is that the estimation of lower and upper bounds of the eigenvalues is not needed. On the other hand, in the SSOR preconditioning by

$$C = \left(\frac{1}{\omega} D + L\right)\left(\frac{1}{\omega} D\right)^{-1}\left(\frac{1}{\omega} D + L^T\right) ,$$

the upper bound $b = 1/(2 - \omega)$ is often a very accurate bound for the largest eigenvalue of $C^{-1}A$ (cf. Andersson [25]). Thus the following combined method may save some computational cost: At first a rough estimate of the smallest eigenvalue is made, an *upper* bound a is to be found! Then the Chebyshev semi-iterative method is applied with parameters determined by a,b (see Section 3.1), until a sufficient decrease of the errors at the eigenvalues in [a,b] is achieved. The errors now look like those in Figure 6.1.

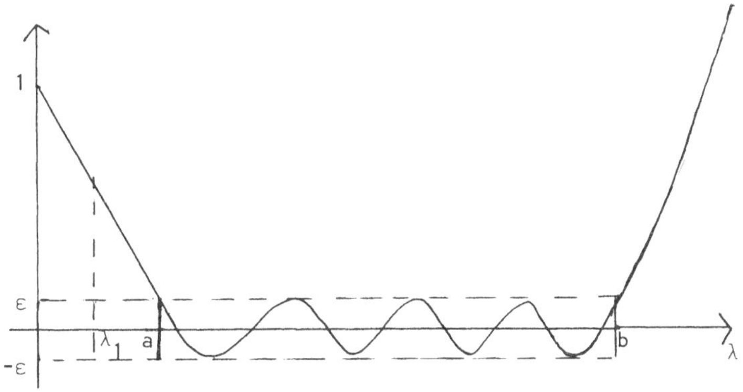

Figure 6.1

In the interval $[\lambda_1, a]$ the errors have decreased by a factor less than

$$2\left(\frac{\sqrt{b-\lambda} + \sqrt{a-\lambda}}{\sqrt{b-\lambda} - \sqrt{a-\lambda}} \; \frac{\sqrt{b} - \sqrt{a}}{\sqrt{b} + \sqrt{a}}\right)^p \sim 2\left(1 - \frac{\nu}{1 + \sqrt{1-\nu}} \sqrt{\frac{a}{b}}\right)^{2p} , \quad \frac{a}{b} \to 0$$

where $\nu = \lambda/a$, after p iterations. This is easily found from the explicit expression of the Chebyshev polynomials. In order to decrease all errors within a relative accuracy of ε, the conjugate gradient method is now applied, and only a few iterations are often needed since they are essentially determined by the condition number $a/\lambda_1 \ll \lambda_0/\lambda_1$. The technique can also be repeated in several (outer) steps; the Chebyshev method can then be regarded as an inner iterative method and the conjugate gradient method as an outer method.

This,or a similar technique (but without preconditioning of A), is used by Rutishauser in [27]; see also Wachspress [32]. It may be mentioned at this point that the so called coarse mesh rebalancing technique can also be viewed upon in this fashion (see [32]).

Finally, we remark that the above technique may also be useful in the calculation of a set of smallest eigenvalues of A (see [27]). Then, first the Chebyshev technique is used for smoothing the initial vector; i.e., for obtaining a vector in a subspace spanned by essentially only small eigenvalue components. To get all of the small components present, it may be advisable to take a linear combination of two or more such vectors using of course different initial vectors r^0. Then Lanczos' method [33] (or the conjugate gradient method) is applied in order to get the corresponding minimal polynomial.

To accelerate the Chebyshev process, preconditioning is advantageous as long as small eigenvalues of A are also small eigenvalues of $C^{-1}A$. The computational complexity in this initial smoothing phase is then negligible compared to the final phase.

Again, it may be necessary to use the method as an outer-inner procedure (cf. the so called cgT method in [27]).

7. Applications of iterative methods to partial differential equation problems.

7.1. History of some iterative methods for differenced Dirichlet problems.

Let

$$L_h u_\alpha = b_\alpha u_\alpha - \sum_{i=1}^{n} c_{\alpha,e_i} u_{\alpha+e_i} + c_{\alpha-e_i,e_i} u_{\alpha-e_i} = \tilde{g}_\alpha , \qquad x_\alpha \in \Omega_h$$

$$u_\alpha = f(x_\alpha) , \qquad x_\alpha \in \partial\Omega_h$$

where α is a multi-index and e_i is the unit vector in the positive i^{th} direction, be the differenced form of the Dirichlet problem

$$Lu = - \sum_i \frac{\partial}{\partial x_i} [a_i(x) \frac{\partial u}{\partial x_i}] + q(x)u = g(x) , \qquad x \in \Omega$$

$$u(x) = f(x) , \qquad x \in \partial\Omega ,$$

on an open bounded domain Ω with boundary $\partial\Omega$, where $a_i(x) > 0$, $q(x) \geq 0$, $x \in \bar{\Omega}$. Let h be a mesh width parameter. In matrix form we write the system of linear equations above as

$$Au = \tilde{f} , \qquad A = D + L + U$$

where $D = \text{diag}(A)$ and L,U are the lower and upper triangular part of A, respectively.

In the particular case $a_i \equiv 1$ (and $q \equiv g \equiv 0$) with L_h defined by the "molecule"

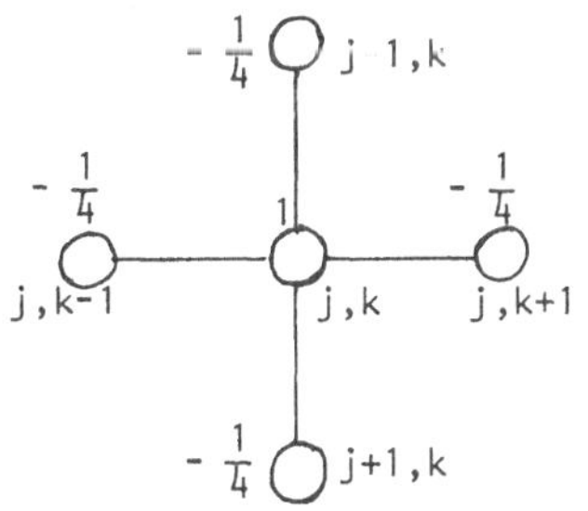

the numerical solution of this problem by iteration was dealt with by Richardson [20] (1911, 1925). Written in the form of a simultaneous overrelaxation method, we have

$$u_{j,k}^{(l+1)} = u_{j,k}^{(l)} - \omega L_h u_{j,k}^{(l)} , \qquad u_{j,k}^{(0)} \text{ arbitrary,}$$

where $\omega > 0$ is a parameter. To decrease the norm of the error vector by a factor $\varepsilon > 0$ demands $O(h^{-2} \log \varepsilon)$ iterations. Liebmann [34] (1918) considered a successive overrelaxation method (SOR) which (with $a_i \equiv 1$, $q \equiv g \equiv 0$) may be written

$$u_{j,k}^{(l+1)} = u_{j,k}^{(l)} - \omega[u_{j,k}^{(l)} - \frac{1}{4}(u_{j-1,k}^{(l+1)} + u_{j+1,k}^{(l)} + u_{j,k-1}^{(l+1)} + u_{j,k+1}^{(l)})]$$

in the row-wise ordering. In matrix form we have

$$(D + \omega L)u^{(l+1)} = [(1 - \omega)D - \omega U]u^{(l)} + \omega\tilde{f} .$$

We observe that this use of "improved" values as soon as they are computed makes the efficiency of the SOR method a function of the ordering of the equations.

If $\omega = 1$, this method is of the form of the method of Gauss-Seidel and then demands approximately half as many iterations as the Richardson method for $\omega = 1$ (which is of the form of the Gauss-Jacobi method). There is a thorough theory of the SOR method, using Perron-Frobenius theory for non-negative square matrices (see Varga [1] (1962) and Young [17] (1971)). For so called consistently ordered equations, there is an optimal value of ω, $1 < \omega < 2$, $\omega_{opt} = 2/\{1 + \sqrt{1 - [\rho(D^{-1}(L + U))]^2}\}$, where $\rho(A)$ is the spectral radius of A. This is valid for all consistent orderings.

For $\omega = \omega_{opt}$, the number of iterations is $O(h^{-1} \log \varepsilon)$. However, this number is sensitive to the choice of ω, and furthermore, since the eigenvalues of the associated matrix of iteration $(D + \omega L)^{-1}[(1 - \omega)D - \omega U]$ are situated on the circle $|z| = \omega_{opt} - 1$ in the complex plane, no acceleration method as given exists for the optimal SOR method (cf. remark made at the end of Section 6.2).

However, Sheldon [35] (1955) generalized a method of Aitken [36] (1950), going alternatively back and forth over the net with the SOR method. Since the corresponding matrix of iteration

$$M = (D + \omega U)^{-1}[(1 - \omega)D - \omega L](D + \omega L)^{-1}[(1 - \omega)D - \omega U]$$

is similar to a symmetric matrix if $U = L^*$, this method is called the symmetric successive overrelaxation method (SSOR method). It is easily proved that if A is positive definite and $0 < \omega < 2$ then the eigenvalues of M are in the interval $[0,1)$; i.e., the method is convergent. The efficiency of this method is dependent on the ordering but only slightly on the parameter ω. If the condition $\rho(\tilde{L}\tilde{L}^*) \leq 1/4$, where $\tilde{L} = D^{-1/2}LD^{-1/2}$, is satisfied, the number of necessary iterations is $O(h^{-1} \log \varepsilon)$. This was proved by Habetler and Wachspress [10] (1961). Ehrlich [37] generalized the SSOR method to a block SSOR method using block-diagonal matrices D.

Since the eigenvalues of M are real we can use an acceleration method based on the Chebyshev polynomials. This was noted by Sheldon [35]. The number of iterations is now only $O(h^{-1/2} \log \varepsilon)$. It is easily seen that the condition $\rho(\widetilde{LL^*}) \leq 1/4$ is satisfied if the coefficients $a_i(x)$ are constant over Ω. If this is not valid, a scaling technique or, equivalently, a technique using one iteration parameter ω per mesh point can be used to obtain the same asymptotic rate of convergence as in the constant coefficient case. This is shown in Dupont et al. [12] and in Axelsson [13].

The Richardson method was generalized by a number of authors, among them Young [38] (1954), using variable parameters ω_l to yield a Chebyshev acceleration with the Chebyshev polynomial T_p after a fixed and predetermined number p of iterations was performed. The straightforward application of this method, however, is numerically unstable, but a two-step version of the method is stable and demands no predetermination of the number of iterations. A two-step version can be found in the paper of Stiefel [39] (1958) (cf. Section 3.2). The conjugate gradient method was applied to partial differential equations in Engeli et al. [27] (1959). Further references have already been given.

7.2. SSOR preconditioning of finite difference and finite element matrices.

In the classical finite difference approximation of a 2nd order partial differential equation problem with constant coefficients, it is seen at once that the condition (5.5) is satisfied for the natural (row-wise) ordering. In this ordering the balancing between predecessor and postdecessors (see Section 5.4) is satisfied as there is in general an equal number of them. On the other hand, in the so called σ_1-ordering (see e.g. [17]) where each second point is numbered, this balancing is violated at its extreme; either there are only coupling to postdecessors or only to predecessors. It is also well known that the SSOR preconditioning in this case works best with $\omega = 1$, the Aitken [36] method. However, there may be other orderings which work as well as the natural. For example, the diagonal and also many Cuthill-McKee orderings [40] are well balanced in the above sense (see [41]).

For differential equation problems with smoothly varying coefficients it is also possible to prove that the (natural ordered) SSOR preconditioning condition *(ii)* in Section 5.4 is satisfied with $\delta = O(1)$, $N \to \infty$. The proof goes as follows:
We have

$$(LD^{-1}L^T x,x) - \frac{1}{4}(Dx,x) \leq \max \lambda = \rho, \qquad \forall x \mid (x,x) = 1 ,$$

where λ is an eigenvalue of $LD^{-1}L^T - \frac{1}{4}D$. Since all terms in $LD^{-1}L^T$ are non-negative in the second-order finite difference approximation, in applying the classical

Gerschgorin theorem it suffices to bound the row sums of the matrix $-\frac{1}{4}D + LD^{-1}L^T$. Thus

$$\rho \leq \max_{\alpha}\left\{-\frac{1}{4}b_\alpha + \sum_{i=1}^{n}\left[c_{\alpha-e_i,e_i}b_{\alpha-e_i}^{-1}\sum_{j=1}^{n}c_{\alpha-e_i,e_j}\right]\right\}.$$

Let

$$\eta_\alpha = \Sigma\, c_{\alpha-e_i,e_i} \,/\, \Sigma\, c_{\alpha,e_i}$$

$$\beta_\alpha = \min_{i=1,\dots,n} \eta_{\alpha-e_i}.$$

Then

$$\rho \leq \max_{\alpha}\left\{\frac{1}{4}(\Sigma\, c_{\alpha,e_i})\left[-1 + \eta_\alpha\left(-1 + 4\frac{1}{1+\beta_\alpha}\right)\right]\right\}.$$

Since $(Ax,x)/(x,x) \geq O(h^2)$, $h \to 0$, it suffices to bound the term in braces from above by Ch^2 in order to show that $\rho \leq (Ax,x)/(x,x)\ \forall x$ (cf. Section 5.4). We rewrite this term as

$$\frac{1}{1+\eta_\alpha}\left[-(1+\eta_\alpha)^2 + 4\eta_\alpha\right] + 4\eta_\alpha\left[\frac{1}{1+\beta_\alpha} - \frac{1}{1+\eta_\alpha}\right]$$

$$= -\frac{(1-\eta_\alpha)^2}{1+\eta_\alpha} + \frac{4\eta_\alpha}{(1+\beta_\alpha)(1+\eta_\alpha)}(\eta_\alpha - \beta_\alpha).$$

For some $i = i_0$ we have

$$\beta_\alpha = \eta_{\alpha-e_{i_0}}.$$

Thus, if a_i have a Lipschitz-continuous first derivative, we have

$$\eta_\alpha = 1 - h\frac{\Sigma\frac{\partial}{\partial x_i}a_i(x_\alpha + \frac{1}{2}e_i)}{\Sigma\, a_i(x_\alpha + \frac{1}{2}e_i)} + O(h^2),$$

and the desired bound of the row sums is achieved.

We remark that it is also possible to satisfy condition (*ii*) in Section 5.4 with $\delta = O(1)$, even if the coefficients $a_i(x)$ in the differential equation are only Lipschitz-continuous (in the e_i direction), if a particular matrix $\tilde{D}$ is used (see Axelsson [13]).

Numerical experiments with problems with discontinuous coefficients have shown only a slight increase in the number of iterations. For average size problems (up to 1000 unknowns) with one line of discontinuity, the increase was at most 50% of the number of iterations for all rates of discontinuity between the coefficients in the two regions.

We will now consider SSOR preconditioning of nearly Toeplitz matrices with application to finite element equations.

Let

$$A = [a_{ij}] , \qquad i,j = \ldots,-1,0,1,\ldots$$

be an infinite Toeplitz matrix; i.e.,

$$\alpha_{ij} = a_{j-i} ,$$

where the numbers a_k thus determine the matrix. The study of finite Toeplitz matrices

$$A^{(n)} = [a_{j-i}] , \qquad i,j = 1,\ldots,n$$

obviously has some relevance to regular finite elements and a theory for SSOR preconditioning of such matrices is of interest since we have no longer necessarily diagonal dominance (cf. Section 5.3). It is possible to show that the SSOR preconditioning of a Toeplitz matrix with spectral condition number $O(n^{2m})$ results in a spectral condition number $O(n^{2m-1})$. This result is sharp and is also valid for block Toeplitz matrices with n being the order of the blocks (see Andersson [25]).

The perturbation of the Toeplitz structure in a discretized partial differential problem due to boundary conditions and non-constant (but smoothly varying) coefficients can also be dealt with (see [25]) giving the same results as above. Numerical tests on problems not covered by theory give evidence of a more general applicability of this preconditioning.

To summarize, the spectral condition number of a finite element matrix achieved by the discretization of a $2m^{th}$ order partial differential equation problem is $O(h^{-2m})$, where h is the size of some average element side (see also Fried [42]). The SSOR preconditioning results in a spectral condition number $O(h^{-2m+1})$. This has only been proved for almost regular problems, however.

There is also an interesting clustering effect which in a two-point boundary value problem results in an eigenvalue distribution of the SSOR preconditioned matrix, where all eigenvalues except one are clustered around 1 and the last is approximately $1/(2-\omega)$ and goes to infinity with $h \to 0$. A more general result for *one*-dimensional Toeplitz matrices is that for such matrices the SSOR preconditioning results in a similar clustering but with a finite number (independent of $n = h^{-1}$) of eigenvalues not belonging to this cluster. The results of Section 4.2 now tell us that the effective spectral condition number is $O(n^{2m-2})$ (see [25]).

A proof of the applicability of preconditioning for finite element matrices may also be based on spectral equivalence: If two sequences of matrices $\{A_N\}$ and $\{B_N\}$, where N is the order of the matrices, satisfy

$$\alpha(B_N x,x) \leq (A_N x,x) \leq \beta(B_N x,x), \qquad x \in C^N, \quad 0 < \alpha,$$

α,β independent of the number of unknowns N, then the two sequences are said to be *spectrally equivalent*. A sequence of preconditioning matrices $\{C_N\}$ which works well for $\{B_N\}$ then also works well for $\{A_N\}$ (cf. [11] and [13]).

7.3. Some special methods.

Here we will briefly mention some methods for some particular partial differential equation problems. Good results were achieved for the coupled method in (6.1) applied to quasi-stationary (periodic) partial differential equations. The matrices A,B in $A + i\omega B$ are positive definite and the number of outer iterations were independent of the condition numbers of A,B. For a relative accuracy of $\varepsilon = 10^{-6}$, 4 to 7 outer iterations were enough. A similar method has been applied to the biharmonic equation problem (see [43]); see also [44], [45], [46] and [47]). Here, however, the number of outer iterations is dependent on the spectral condition number.

For nonsymmetric equations, the Chebyshev method for matrices with complex eigenvalues (see Section 6.3) has been applied (see [48]).

Finally, we mention that good results have been achieved for Fedorenko type [49] methods for elliptic problems with smoothing based on interpolation from courser to finer grids (see e.g. [50]).

7.4. Some numerical results for the SSOR preconditioned conjugate gradient method.

In this final section we will give some numerical results for 2nd order problems and make comparisons with some other methods. One reason for not dealing with higher order problems is that the best iterative scheme, which is likely of the coupled type (cf. the previous section), has perhaps not yet been devised for 4'th order (and higher order) problems. According to the theory of the SSOR preconditioning, the number of iterations is $O(h^{-m+1/2})$, whereas there are already certain coupled schemes which only require $O(h^{-1})$ iterations for the biharmonic problem ($m = 2$).

The theory for the SSOR preconditioning of finite element methods covers only regular or almost regular elements (see Section 7.2). Numerical experiments, including high-order approximations and almost uniform elements, indicate however the advantages of the preconditioned iterative technique also for problems not yet covered by theory.

In the following table we show the number of arithmetic operations for a self-adjoint 2nd order problem in a d-dimensional space with smooth enough coefficients. N is the number of nodes.

Table 7.1

Number of arithmetic operations

Method	Preconditioning work	Work per iterative step	Number of iterations	Iterative work (total)	Storage
Conjugate gradient	0	$(c+5)N$	$\frac{1}{2}\sqrt{\mathcal{H}(A)}\ln(2/\varepsilon)$	$O(N^{1+1/d})\ln N$	$(c+4)N$
SSOR precond. conjugate gradient	0	$2(c+3)N$	$\frac{1}{2}\sqrt{\mathcal{H}^*}\ln(2/\varepsilon)$	$O(N^{1+1/d})\ln N$	$(c+5)N$
Cholesky factorization (natural ordering)	$O(N^{3-2/d})$	$O(N^{2-1/d})$	$O(1)$	$O(N^{2-1/d})$	$O(N^{2-1/d})$

Remark 1. In the Cholesky factorization method the "preconditioning work" is that spent for the factorization itself, and "the work per iterative step" is that spent for the forward and backward substitutions.

Remark 2. In computing the total iterative work we have supposed that the relative accuracy is $\varepsilon = O(N^{-\mu})$, $\mu > 0$.

Remark 3. c is the average number of non-zero elements in each row of A.

Remark 4. $\mathcal{H}^*$ is the effective spectral condition number of the preconditioned matrix. If d = 1, this number is actually only O(1).

If the equation is to be solved for many right hand sides b with the same finite element matrix A, it is probably worthwile to do some preconditioning work (cf. Section 5.3 where some more accurate approximate factorization techniques are mentioned).

Since for large enough problems in at least d = 2 dimensions the iterative work is larger for the Cholesky method than for the SSOR preconditioned conjugate gradient method, there is however no reason to use the former method. If d = 3 even the pure conjugate gradient method is faster for large enough problems. Also, it is clear that there must be a large number $O(N^{1/d})$ of right hand sides, else the preconditioning work (the factorization) will be dominant anyway.

If memory requirements are taken into account, it is even more clear that it is uneconomical to use Cholesky factorization with the natural (row by row) ordering.

In [52] and [53] other orderings are dealt with for plane meshes. The ordering which in connection with Cholesky factorization minimizes (in the asymptotic sense) both the number of arithmetic operations and fill-in (of non-zero elements) is the so-called nested dissection method. For plane regular problems the arithmetic is $O(N^{1.5})$ and the fill suffered during the factorization is $O(N \ln N)$. This method can stand comparison with the SSOR preconditioned conjugate gradient method (for which the arithmetic for a plane problem is $O(N^{1.25})$ if there are enough right hand sides, even for large problems. This ordering is quite complicated,however, and the implementation needs a skillfull programmer who can avoid prohibitive overhead operations.

In [53] a simpler (one-way) dissection ordering is also presented for which the arithmetic and fill-in is $O(N^{1.75})$ and $O(N^{1.25})$, respectively. Once again, if there are many right hand sides this method can stand comparison with the preconditioned conjugate gradient method. From the above follows only the asymptotic behaviour. It is thus of interest to compare the actual number of operations using different methods for some average sized problems. In [53] Alan George has given figures for the model 5-point difference approximation scheme on a square mesh. Interestingly enough, the solution time for the nested dissection schemes is not much shorter than for the natural ordering for problems of average size, up to say N = 1600. At best the number of operations in the one-way dissection ordering (which is actually less than those for the full dissection ordering for $N \leq 4000$) is half that of the natural ordering scheme. The factorization time when N = 1600 for the dissection ordering is about 1/3 of that of the natural ordering as well as of the one-way dissection ordering.

If the preconditioned conjugate gradient method is applied to the same problem (N = 1600), but without taking advantage of the regularity of the element matrices, the number of multiplications (and additions) to reach a relative accuracy of 10^{-3} is roughly 195 000. This is 1/8 of the total number of multiplications for Cholesky factorization with natural ordering and only 3 times more than what is needed for the solution time in the fastest ordering. It is necessary to have 22 right hand sides in order to make the number of operations in the natural ordering equal to that of the above iterative scheme.

In the above considerations overhead operations and storage requirements have been neglected. Had they been considered the iterative method might have been still faster than the direct methods. The storage needed is actually only 1/2 of what is needed in the best one-way dissection scheme and less than 1/4 of that in the natural ordering scheme. Again, in these considerations no use of the regularity of the mesh has been utilized in the iterative method.

If the differences in solution times are perhaps not large in the above plane problem, they will certainly be much larger in a 3-dimensional problem; all the more as the merits of the dissection ordering have been established for plane problems only. This author therefore believes that for *3-dimensional* problems no ordering exists for a general region which can stand comparison with the preconditioned conjugate gradient method, even if only the number of arithmetic operations is compared. That the band Cholesky factorization method is too expensive for 3-dimensional problems is clear from Table 7.1. The factorization time is $O(N^{\alpha}), \alpha = 7/3$, whereas the total time for the SSOR iterative conjugate gradient method is only $O(N^{\alpha/2})$.

Other advantages with iterative methods are that they are easy to program and have minimum storage requirements so it is possible to stay in core much longer than would be possible with a Cholesky factorization method. If the differential equation coefficients are constant over a region where regular elements are chosen, only the coefficients for one finite element matrix in this region have to be stored. Furthermore, iterative methods lend themselves naturally to the case where there is a good initial approximation to the numerical solution being calculated. Situations where good initial approximations are given occur for instance in time-dependent problems like heat conduction, in interactive design of boundary value problems and in quasilinear problems.

In some problems it may be advantageous not to *assemble* the stiffness matrix A. Such a situation occurs when, due to the cancellation of digits during the assembling of A, the coefficients of A have large relative errors. In an iterative method, where only the residual vectors are needed, this can be avoided by calculating the residual as the sum over the element matrices k_i,

$$r^l = \sum_{i=1}^{N'} a_i^T k_i (a_i u^l) - b ,$$

where N' is the number of elements and a_i is the portion of the Boolean connectivity matrix corresponding to the i'th element. This technique is mentioned e.g. in [54] and [42]. Actually, this will also decrease the rounding errors during the iterative steps. This was noted in a simple (2 point) boundary value problem in [55].

Furthermore, in an iterative method it is not necessary to calculate more digits of accuracy than are wanted. The relative accuracy ε can be chosen so small that it is of the same size as the inherent relative error in the right-hand side element b, including the discretization error. This means that in practice, single precision calculation suffices on most computers.

Let us now compare the iterative procedure with the band (natural ordering) Cholesky method for very large problems where also the iterative method needs backing store. Let q be the number of data transfers between the secondary and the main storage which take place. In the Cholesky method we neglect the number of data transfers during the factorization. Thus only the solution phase (backward and forward substitutions) is considered in this method. Then for large enough problems it is easily seen that q is directly proportional to the number of arithmetic operations in both methods. Thus, if the SSOR condition is satisfied, we see from Table 7.1 that the solution phase needs $O(N^{1-3/2d}/\ln N)$ times more data transfers than the SSOR preconditioned conjugate gradient method. For the conjugate gradient method alone the corresponding relation is $O(N^{1-2/d})$; i.e., for large 3-dimensional problems even the pure conjugate gradient method needs a fewer number of data transfers.

Let us finally remark that successful attempts (using the preconditioned conjugate gradient method) have been made for problems with discontinuous and nonlinear material coefficients.

7.5. Conclusions.

In this final section we summarize the results achieved so far for finite element problems and make a comparison with direct solvers. Except for problems close to model problems (either smooth coefficients or simple geometry) the results are based on numerical evidence only. The results are summarized in Table 7.2.

We thus conclude that in a 3-dimensional 2'nd order problem the iterative methods are even faster (asymptotically) than the pure solution part in the direct solvers.

For a 4'th order plane problem a dissection ordered solver is superior to the iterative solver. In other problems, plane 2'nd order problems or 3-dimensional 4'th order problems, the superiority of one method over another is not clear and the size of the problem and the number of right hand sides should influence the choice of method. The band Cholesky method is not competitive, however, except for small problems.

Table 7.2

State-of-the-art of some linear solvers for discretized 2m'th order self-adjoint elliptic problems.

a. Number of arithmetic operations.

Space dim. → / Order 2m ↓	2	3	Method
2	$O(N^{1.5})$ $O(N^{1.25})$	$O(N^{1.33})$ $O(N^{1.17})$	Conjugate gradient SSOR precond. conj. grad.
	$O(N^2)$ $O(N^{1.5})$ $O(N^{1.5})$ $O(N \log N)$	$O(N^{2.33})$ $O(N^{1.67})$ $O(N^2)$ $O(N^{1.33})$	Cholesky (natural ordering): fact. solution Cholesky (dissection ordering): fact. solution
4	$O(N^2)$ $O(N^{1.67})$	$O(N^{1.67})$ $O(N^{1.5})$	Conjugate gradient SSOR precond. conj. grad.
	As for m = 1	As for m = 1	Cholesky (natural ordering (fact., sol.) and dissection ordering)

b. Storage

Space dim. → / Order 2m ↓	2	3	Method
2,4	$O(N)$	$O(N)$	Conjugate gradient SSOR precond. conj. grad.
	$O(N^{1.5})$ $O(N \log N)$	$O(N^{1.67})$ $O(N^{1.33})$	Cholesky (natural ordering) Cholesky (dissection ordering)

REFERENCES

[1] R.S. Varga, Matrix iterative analysis, Prentice-Hall, New Jersey, 1962.

[2] M.R. Hestenes, Historical papers: Iterative methods for solving linear equations, JOTA 11 (1973), 323-334 *and* The solution of linear equations by minimization, JOTA 11 (1973), 335-359. Completed on December 12, 1951. The first paper originally appeared as NAML Report No. 52-9, 1951.

[3] M.R. Hestenes and E. Stiefel, Methods of conjugate gradients for solving linear systems, J. Res. Nat. Bur. Stand. B49 (1952), 409-436.

[4] R. Fletcher and C.M. Reeves, Function minimization by conjugate gradients, Computer Journal 7 (1964), 149-154.

[5] E. Polak et G. Ribiére, Note sur la convergence de methodes de directions conjuguees, R. I. R. O. 16-R1 (1969), 35-43.

[6] J.W. Daniel, The conjugate gradient method for linear and nonlinear operator equations, SIAM J. Numer. Anal. 4 (1967), 10-26.

[7] R. Bartels and J.W. Daniel, A conjugate gradient approach to nonlinear elliptic boundary value problems in irregular regions, CNA 63, The University of Texas at Austin, 1973.

[8] D.J. Evans, The use of preconditioning in iterative methods for solving linear equations with symmetric positive definite matrices, J. Inst. Maths. Applics. (1968) 4, 295-314.

[9] E.G. D'Yakonov, On an iterative method for the solution of finite difference equations, Dokl. Akad. Nauk SSSR, 138 (1961), 522-525.

[10] G.J. Habetler and E.L. Wachspress, Symmetric successive overrelaxation in solving diffusion difference equations, Math. Comp. 15 (1961), 356-362.

[11] J.E. Gunn, The solution of elliptic difference equations by semi-explicit iterative techniques, SIAM J. Numer. Anal. Ser. B 2 (1964), 24-45.

[12] T. Dupont, R.P. Kendall and H.H. Rachford, Jr., An approximate factorization procedure for solving self-adjoint elliptic difference equations, SIAM J. Numer. Anal. 5 (1968), 559-573.

[13] O. Axelsson, A generalized SSOR method, BIT 13 (1972), 443-467.

[14] O. Axelsson, Iterative methods for elliptic problems, CERN DD/72/13, Geneva, 1972.

[15] O. Axelsson, On preconditioning and convergence acceleration in sparse matrix problems, CERN 74-10, Geneva, 1974.

[16] O. Axelsson, A class of iterative methods for finite element equations, Computer methods in applied mechanics and engineering 7 (1976) (to appear).

[17] D.M. Young, Iterative solution of large linear systems, Academic Press, 1971.

[18] M.J.D. Powell, Restart procedures for the conjugate gradient method, C.S.S. 24 (1975), Harwell, England.

[19] J.K. Reid, On the method of conjugate gradients for the solution of large sparse systems of linear equations, *in* Large sparse sets of linear equations (ed. Reid), Academic Press, 1971.

[20a] L.F. Richardson, The approximate arithmetical solution by finite differences of physical problems involving differential equations, with an application to the stresses in a masonry dam, Trans. Roy. Soc. London A210 (1911), 307-357;

[20b] L.F. Richardson, How to solve differential equations approximately by arithmetic, Math. Gazette, 12 (1925), 415-421.

[21] V.I. Lebedev, S.A. Finogenov, On the order of choice of the iteration parameters in the Chebyshev cyclic iteration method, Zh. Vychislit. Mat. i Mat. Fiz. 11 (1971), 425 and 13 (1973), 18.

[22] O. Axelsson, Lecture notes on Iterative methods, Report 72.04, Department of Computer Sciences, Chalmers University of Technology, Göteborg, Sweden.

[23] D.M. Young, Second degree iterative methods for the solution of large linear systems, J. Approx. Theory 5 (1972), 137.

[24] V.I. Lebedev, Iterative methods for the solution of operator equations with their spectrum on several intervals, Zh. Vychislit. Mat. i Mat. Fiz. 9 (1969), 1247-1252.

[25] L. Andersson, SSOR preconditioning of Toeplitz matrices, Thesis, Chalmers University of Technology, Göteborg, Sweden, 1976.

[26] J. Todd, Introduction to the constructive theory of functions, ISNM vol. 1, Birkhäuser, Basel, 1963.

[27] M. Engeli, Th. Ginsburg, H. Rutishauser, E. Stiefel, Refined iterative methods for computation of the solution and the eigenvalues of self-adjoint boundary value problems, Mitteilungen aus dem Institut für angewandte Mathematik, nr. 8, ETH, Zürich, 1959.

[28] H. Wozniakowski, Numerical stability of iterations for solution of nonlinear equations and large linear systems, Department of Computer Science, Carnegie - Mellon University, 1975.

[29] G. Forsythe, E.G. Straus, On best conditioned matrices, Proc. Amer. Math. Soc. 6 (1955), 340-345.

[30] J.A. Meijerink, H.A. van der Vorst, An iterative solution method for linear systems of which the coefficient matrix is a symmetric M-matrix, TR-1, Academic Computer Centre, Utrecht, the Netherlands, 1976.

[31] P. Concus, G.H. Golub, Use of fast direct methods for the efficient numerical solution of nonseparable elliptic equations, STAN-CS-72-278, Stanford University, 1972.

[32] E.L. Wachspress, Iterative solution of elliptic systems and applications to the neutron diffusion equations of reactor physics, Prentice-Hall, 1966.

[33] C. Lanczos, An iteration method for the solution of the eigenvalue problem of linear differential and integral operators, J. Res. Nat. Bur. Stand. B 45 (1950), 255-282.

[34] H. Liebmann, Die angenäherte Ermittelung harmonischer Funktionen und konformer Abbildungen, Bayer. Akad. Wiss., Math-Phys. Klasse, Sitz. (1918), 385-416.

[35] J.W. Sheldon, On the numerical solution of elliptic equations, Math. Tables and other Aids to Comp. 9 (1955), 101-111.

[36] A.C. Aitken, On the iterative solution of a system of linear equations, Proc. Roy. Soc. Edinburgh A63 (1950), 52.

[37] L.W. Ehrlich, The block symmetric successive overrelaxation method, J. Soc. Indust. Appl. Math. 12 (1964), 807-826.

[38] D. Young, On Richardson's method for solving linear systems with positive definite matrices, J. Math. Phys. 32 (1954), 243-255.

[39] E. Stiefel, Kernel polynomials in linear algebra and their numerical applications, NBS, Appl. Math. Series 49 (1958), 1-22.

[40] E. Cuthill, J. Mc Kee, Reducing the bandwidth of sparse symmetric matrices, Proc. 24th Nat. Conf. of the ACM, ACM Publ. P-69, New York, (1969), 157-172.

[41] W-H. Liu, A.H. Sherman, Comparative analysis of the Cuthill-Mc Kee and the reverse Cuthill-Mc Kee ordering algorithms for sparse matrices, SIAM J. Numer. Anal. 13 (1976), 198-213.

[42] I. Fried, More on gradient iterative methods in finite-element analysis, AIAA Journal 7 (1969), 565-567.

[43] J. Smith, The coupled equation approach to the numerical solution of the Biharmonic equation by finite differences I, II, SIAM J. Numer. Anal. 5 (1968), 323-339, 7 (1970), 104-111.

[44] O. Axelsson, Notes on the numerical solution of the Biharmonic equation, J. Inst. Maths. Applics. 11 (1973), 213-226.

[45] L.W. Ehrlich, Solving the Biharmonic equation as coupled finite difference equations, SIAM J. Numer. Anal. 8 (1971), 278-287.

[46] J.W. Mc Laurin, A general coupled equation approach for solving the Biharmonic boundary value problem, SIAM J. Numer. Anal. 11 (1974), 14-23.

[47] M.M. Gupta, Discretization error estimates for certain splitting procedures for solving first Biharmonic boundary value problems, SIAM J. Numer. Anal. 12 (1975), 364-377.

[48] T.A. Manteuffel, An iterative method for solving nonsymmetric linear systems with dynamic estimation of parameters, UIUCDCS-R-75-758, University of Illinois at Urbana-Champaign, Illinois, 1975.

[49] P.P. Fedorenko, The speed of convergence of one iterative method, Zh. Vychislit. Mat. Fiz. 4 (1964), 559-564.

[50] P.O. Fredrickson, Fast approximate inversion of large sparse linear systems, Mathematics Report # 7-75, Lakehead University, 1975.

[51] P. Concus, G.H. Golub, D.P. O'Leary, A generalized conjugate gradient method for the numerical solution of elliptic partial differential equations, STAN-CS-76-533, Stanford University, 1976.

[52] A. George, Nested dissection of a regular finite element mesh, SIAM J. Numer. Anal. 10 (1973), 345-363.

[53] A. George, Numerical experiments using dissection methods to solve n by n grid problems, Research Report CS-75-07, University of Waterloo, Canada.

[54] R.L. Fox, E.L. Stanton, Developments in structured analysis by direct energy minimization, AIAA Journal 6 (1968), 1036-1042.

[55] P. Pohl, Iterative improvement without double precision in a boundary value problem, BIT 14 (1974), 361-365.

SOLUTION OF LINEAR SYSTEMS OF EQUATIONS: DIRECT METHODS FOR FINITE ELEMENT PROBLEMS

J. Alan George

Dept. of Applied Analysis and Computer Science
University of Waterloo
Waterloo, Ontario, Canada N2L 3G1

CONTENTS

Research supported in part by the National Research Council of Canada under Grant A-8111.

1. Introduction

1.1 Outline and Scope of the Paper

The finite element method is a powerful and highly successful numerical procedure which has been used to obtain approximate solutions to a wide variety of problems in mathematics and engineering [29,32]. The application of the method typically leads to the problem of solving one or perhaps many large sparse systems of linear equations. The general method which is usually used to solve these problems is well known; namely Gaussian elimination. However, since the matrix problems are large and sparse, and because the cost of solving these equations often represents a significant fraction of the total cost of applying the finite element method, numerous techniques and strategies have been and are currently being developed to make Gaussian elimination more efficient for these matrix problems.

However, it should be immediately emphasized that this article is not intended to be a survey of such methods and strategies. For surveys, the reader is referred to [18,26]. Instead, in this article we have selected three ordering and solution schemes to describe in detail, including information about storage schemes, performance of actual computer implementations, and some numerical experiments. The selection of the methods has been largely influenced by the author's own work and the availability of computer programs to produce the numerical experiments. The main objective of this article is to describe several different approaches to the solution of finite element systems of equations. In doing so, we illustrate the interaction between the ordering algorithms and the data structures used to store the sparse matrices, and illustrate the importance of considering the implementation issues of storage and execution overhead when comparing solution methods.

When Gaussian elimination is applied to a sparse matrix A, it usually suffers fill; that is, its triangular factors typically have nonzeros in positions which are zero in A. It is well known that a permutation PAP^T of the rows and columns of A can often significantly change the amount and location of fill, and/or significantly change the amount of arithmetic required for the factorization, assuming that the zeros are exploited. The techniques described in sections 3-6 can be viewed as finding choices for P which meet certain objectives associated with computation and data representation.

In this article we assume A is symmetric and positive definite. This is important because for such matrices PAP^T always has a triangular factorization LL^T, where L is lower triangular. The Cholesky algorithm for computing L is stable [31], which means we have the option of choosing P without concern for numerical stability. Although some finite element formulations lead to unsymmetric problems, the structure of A remains symmetric, so the extension of the ideas in this article to unsymmetric A is in principle quite straightforward. However, one must assume that PAP^T has a triangular factorization for the particular ordering chosen, and that an acceptably accurate factorization is obtained when Gaussian elimination without row and column interchanges is applied.

An outline of the paper is as follows. The remainder of this section contains a characterization of the matrix problems we consider, and a description of our numerical experiments and test problems. Section 2 contains a brief review of some graph theory notions which are used in later sections to describe matrix ordering algorithms. Some useful data structures for the computer representation of graphs are described, along with a recent important algorithm for finding so-called pseudo-peripheral nodes, due to Gibbs et al. [21].

Section 3 contains a modified version of the well known Cuthill-McKee ordering algorithm, along with a description of the profile or envelope storage scheme due to Jennings [23], which is highly appropriate for use with orderings of the type provided by the Cuthill-McKee algorithm. The storage scheme due to Jennings is well known, and is included only for completeness. The Cuthill-McKee algorithm and some of its modifications that we describe are also fairly well known, but the use of the algorithm of Gibbs et al. described in section 2 to find a starting node is a new modification which appears to provide a solution to the one outstanding weakness of the algorithm.

Section 4 is an essential precursor to section 5, and contains some fundamental observations about computations involving partitioned matrices. Two efficient storage schemes for partitioned matrices, which are used in connection with the algorithms of sections 5 and 6, are described. Section 5 is a description of the role that the class of graphs called trees plays in the solution of sparse linear systems. An algorithm for finding a so-called tree partitioning of finite element graphs is described, and this forms the basis for an ordering algorithm for sparse matrix problems. This algorithm combined with the first of the storage schemes of section 4 provides a method for solving finite element matrix problems. Some numerical experiments comparing this approach to the method of section 3 are provided.

Section 6 begins with a brief description of the technique known as substructuring or dissection, and reviews the standard applications of the method. We observe that the recursive or nested application of dissection in the past has not been regarded as a method for finding low-fill or low arithmetic orderings, even though such orderings can usually be obtained this way. A heuristic algorithm is then described for generating a nested dissection ordering and partitioning of general finite element graphs. This algorithm is combined with the second storage scheme of section 4 to provide a method for solving systems of finite element equations. Numerical experiments are provided which suggest the asymptotic behaviour of the algorithm, and a comparison with the method of section 3 is also provided. Section 7 contains our concluding remarks.

1.2 Structure of Finite Element Equations

The application of the finite element method begins with the subdivision of the structure, or region of interest, into smaller pieces called elements. When the object under study is a structure, the elements typically correspond to beams, plates, etc., whereas if the application involves a two dimensional continuum, the domain R is usually divided into triangular or quadrilateral elements, with adjacent elements sharing a common side. For purposes of this discussion, we assume the latter situation, so we have a finite element mesh.

On each element E_j we construct a polynomial characterized by its nodal values x_i at certain node points on E_j, typically at its vertices, and perhaps also at points on its edges and interior. For example, a cubic polynomial in two variables can be uniquely characterized on a triangle by values at its vertices, points of trisection of its sides, and at its centroid. It can alternatively be characterized by values and first derivatives at the vertices, and the value at the centroid. The term "element" is often used to denote a particular polynomial-geometry pair, as well as the physical subdivisions of the domain. For example, one often sees references to "quadratic triangular elements" or "linear quadrilateral elements" in the literature. Descriptions of various elements can be found in [32].

Constructing a polynomial for each element produces a piecewise polynomial $v(x,y)$ defined on R, and since adjacent elements have common nodal values, $v(x,y)$ will have a certain degree of continuity automatically imposed across interelement boundaries. It is easy to verify that this procedure amounts to constructing a set of basis functions ϕ_i, one for each nodal value x_i, whose support is limited to elements sharing the corresponding node, as shown in Figure 1.2.1. The piecewise polynomial $v(x,y)$ thus can be expressed in the form $v(x,y) = \sum_{i=1}^{N} x_i\phi_i$, where N is the total number of nodal values.

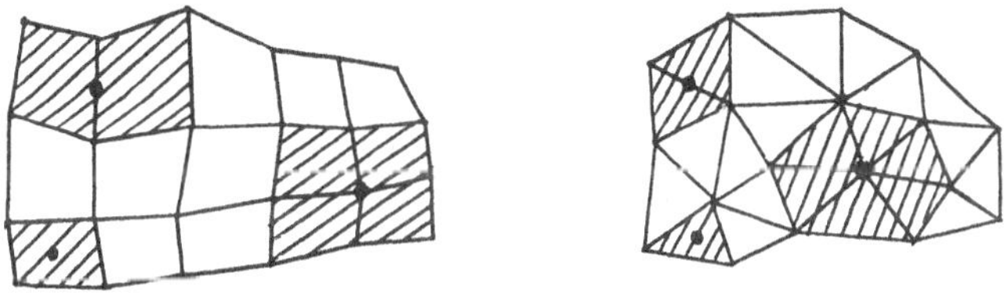

Figure 1.2.1 Support of basis functions associated with vertex, edge and interior nodes on triangular and quadrilateral meshes

The final step of the finite element method is to determine the nodal values x_i; this selection is carried out according to some variational or orthogonality principle, such as the well-known Rayleigh-Ritz or Galerkin procedures, yielding a system of linear algebraic equations.

Suppose for definiteness, we have the two-dimensional variational problem

$$F(u) = \iint_R [a_1 u_x^2 + 2a_2 u_x u_y + a_3 u_y^2 - 2fu]dxdy = \min \tag{1.2.1}$$

$$u = g \text{ on } \partial R,$$

where R is some polygonal domain such as shown in Figure 1.2.1, ∂R is its boundary, and $a_1(x,y)a_3(x,y) > a_2(x,y)^2$ in R. Substituting $v(x,y)$ into (1.2.1) and minimizing with respect to the x_i's yields the system

$$Ax = b, \tag{1.2.2}$$

where

$$A_{ij} = \iint_R [a_1 \frac{\partial\phi_i}{\partial x}\frac{\partial\phi_j}{\partial x} + a_2(\frac{\partial\phi_i}{\partial x}\frac{\partial\phi_j}{\partial y} + \frac{\partial\phi_i}{\partial y}\frac{\partial\phi_j}{\partial x}) + a_3 \frac{\partial\phi_i}{\partial y}\frac{\partial\phi_j}{\partial y}]dxdy \tag{1.2.3}$$

and

$$b_i = \iint_R f\phi_i dxdy. \tag{1.2.4}$$

From (1.2.3) it is clear that A will be sparse; obviously $A_{ij} = 0$ unless ϕ_i and ϕ_j have overlapping support, and this will happen only if x_i and x_j are nodal values associated with the same element.

In practice, boundary conditions will determine some of the nodal values, implying that some of the equations of (1.2.2) are replaced by trivial equations of the form $x_i = b_i$. In theory such equations could be immediately eliminated (thus reducing the size of the system), but most finite element programs solve the expanded system, rather than rearranging the equations after the boundary conditions have been imposed. This results in simpler coding, and provides more freedom in selecting the point in the equation generation process where boundary conditions are imposed. For our test problems, described in section 1.3, we simply generate various triangulations of some two-dimensional domains, and then construct a positive definite matrix having the appropriate finite element structure. The influence of any boundary conditions that might occur in an actual physical problem of the same shape is ignored. In any case, for large N the number of nodal values prescribed by boundary conditions is typically relatively small.

1.3 Test Problems and Remarks on the Numerical Experiments

One of the objectives of this paper is to provide the reader with information regarding the effectiveness of various ordering and solution strategies for solving finite element equations. In order to do this, we have provided details on the computer implementation of the several algorithms we discuss, and we report on the performance of these computer programs. In this section we describe our test examples on which our experiments were performed.

Our test problems are positive definite matrix equations arising from 6 two-parameter finite element meshes typical of those that might arise in structural analysis or the study of heat conduction. The basic meshes, shown in Figure 1.3.1,

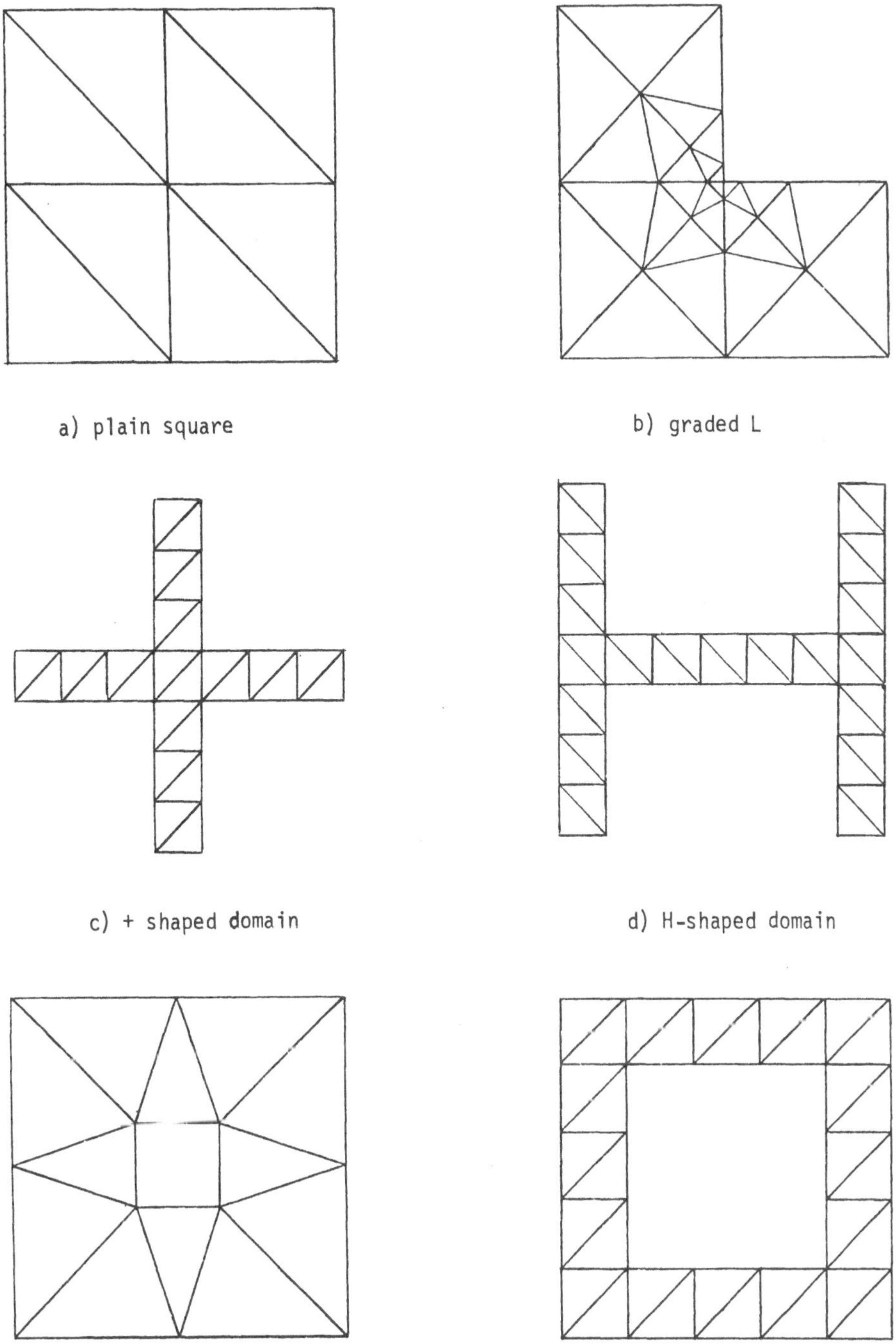

a) plain square

b) graded L

c) + shaped domain

d) H-shaped domain

e) square (small hole)

f) square (large hole)

Figure 1.3.1 Mesh problems with s=1

are subdivided by a factor s in the obvious way, yielding a mesh having s^2 as many elements as the original, as shown in Figure 1.3.2 for the "+" shaped domain with s = 6. The second parameter t governs the distribution of nodes in the mesh, and

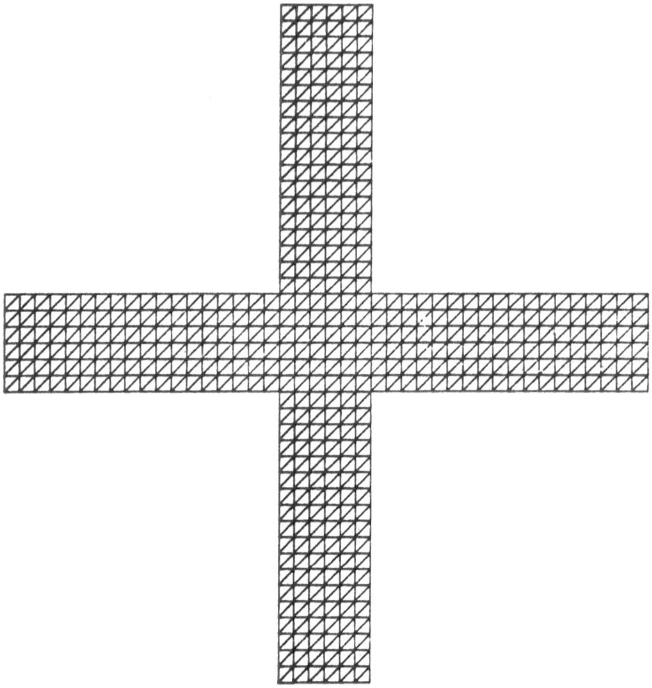

Figure 1.3.2 + shaped domain with s=6

essentially corresponds to the degree of the so-called triangular elements [32,p.115]. For $t \geq 1$, there is one node at each triangle vertex, t-1 nodes along each triangle side, and (t-1)(t-2)/2 nodes in each triangle interior. Figure 1.3.3 displays the node distribution for t = 1,2,3,4.

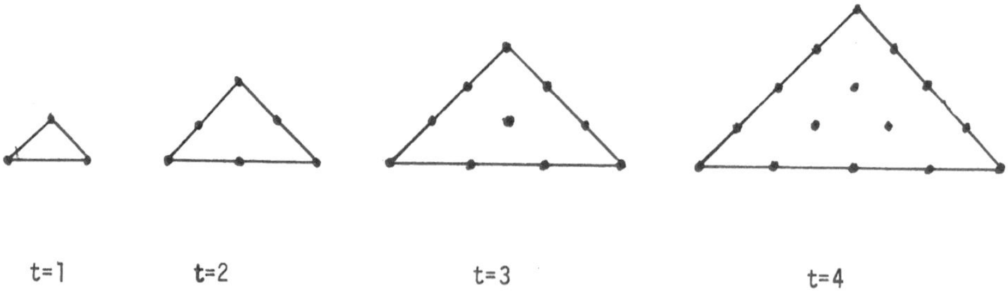

Figure 1.3.3 Element node distribution for $1 \leq t \leq 4$.

Providing a basic mesh along with s and t determines a specific finite element mesh with N nodes. For some labelling of the nodes, we generate a positive definite N by N matrix problem $Ax = b$ where $A_{ij} \neq 0$ if and only if nodal values x_i and x_j are associated with nodes of the same element. The graph G(A) of such a matrix is referred to as a finite element graph.

Although our test problems are restricted to two dimensions, and to a certain class of element, some independent experiments for other classes of problems have provided results similar to the ones we report in this article. We emphasize that our algorithms and programs make no use whatsoever of the fact that the test problems are of this special class.

In our numerical experiments we include both primary storage and overhead storage; that is, the storage used for actual numerical values and that used for pointers etc. associated with the matrix sparsity structure. Since one pays for storage regardless of the way it is employed, it is unfair in any practical study to ignore the overhead component. We do however distinguish between the two uses because on some computers having a large word size, it may be sensible to "pack" two or more pointers into each word. Our experiments were performed on an IBM 360/75 in single precision, and both pointers and the matrix entries were stored using 32 bits; however, by reporting primary and overhead storage separately, the reader can gauge their relative importance on other computers.

In our numerical experiments we also separate out 1) the execution time due to the actual determination of the ordering, 2) the time required for the factorization of A into LL^T, and 3) the time required for the triangular solution given the factorization (i.e., solution of the triangular systems $Ly = b$ and $L^Tx = y$). We do this for two reasons. First, in some situations, many different problems having the same zero-nonzero structure must be solved. In this case, it may be quite reasonable to ignore the ordering and initialization time in comparing methods. Second, in the solution of some mildly nonlinear or time dependent problems, many systems having the same coefficient matrix but different right hand sides must be solved. In this context, the cost of solving the problem, given the ordering and the factorization, may be the primary factor in comparing methods.

In practical terms, one is usually interested in comparing methods so as to determine the method which will yield the lowest computer charges. How does one measure cost? Clearly the real cost C is some function of execution time T and storage used S. Since the use of one method rather than another may simply increase T and reduce S, the function C(S,T) will be fundamental in determining which method results in the lowest charges. Computer memory continues to be a relatively expensive resource, so computer charging algorithms are usually designed to discourage large storage demands. Typically, C(S,T) is of the form $T \times p(S)$, where p is a polynomial of degree ≥ 1. Thus, we contend that storage reduction should be regarded as at least as important as reduction in execution time. It is frequently the case that a method which doubles the execution time and halves the storage requirements of another method yields a net reduction in computer charges. Another important related point is that a reduced storage requirement may allow one to avoid using auxiliary storage. The value of this is hard to assess quantitatively, but the cost of data transfer can be substantial.

The main point of these remarks about charging is to demonstrate that the determination of which method is best in the ultimate monetary sense is almost certainly a function of ones computing environment. Since some schemes we discuss in this article tend to decrease storage at the expense of increased execution time, the reader should not expect firm conclusions or recommendations. To a large extent we simply present experimental results which illustrate the issues we have just raised about the computer solution of sparse linear systems of equations.

In the various tables which appear in the sequel, operations mean multiplications and divisions. We sometimes report both execution times and operation counts to confirm that the operation counts are an accurate reflection of the amount of computation being performed. Since sparse matrix computations often involve rather sophisticated data structures, one must be assured that the computer time associated with data structure manipulation is at worst proportional to the amount of arithmetic being performed. Otherwise, comparing methods on the basis of operation counts makes no practical sense. We will expand on these points when we report our numerical experiments in sections 3-6. All execution times are reported in seconds on an IBM 360/75 computer. The times reported have the usual errors in them due to the inevitable memory interference and cycle stealing which occurs on modern multi-programmed computing systems, but we have tried to minimize the errors as much as possible by making multiple runs, and running programs when the computer was lightly loaded.

We refer to the same set of test examples in several different tables, and it is helpful to abbreviate these references through use of the following table.

TABLE 1.3.1

Test Examples

Test Example	Domain	Degree t	No. of Division s	No. of Equations N
a.1	Square	1	16	1089
a.2	Square	2	8	1089
b.1	Graded L	1	8	1009
b.2	Graded L	2	4	1009
c.1	+ shaped	1	9	1180
c.2	+ shaped	2	4	945
d.1	H shaped	1	8	1377
d.2	H shaped	2	3	805
e.1	Square (small hole)	1	12	936
e.2	Square (small hole)	2	6	936
f.1	Square (large hole)	1	9	1440
f.2	Square (large hole)	2	4	1152

2. Graph Theory

2.1 Introduction

In this section we review some basic graph theory ideas which are useful in the description of ordering algorithms for sparse matrices. Although rather few results from graph theory have found direct application to the analysis of sparse matrix computations, the language and notation is convenient and helpful in describing algorithms. In addition to introducing some basic definitions, we also describe some data structures for the computer representation of graphs, along with a recently developed algorithm for finding so-called pseudo-peripheral nodes [21]. This algorithm is used in all three ordering algorithms discussed in this article.

2.2 Basic Terminology and Some Definitions

An undirected graph $G = (X,E)$ consists of a finite set X of nodes or vertices together with a set E of edges, which are unordered pairs of distinct nodes of X. A subgraph $G' = (X',E')$ of G is one for which $X' \subseteq X$ and $E' \subset E$. For $Y \subset X$, the section graph $G(Y)$ is the subgraph $(Y,E(Y))$, where

(2.2.1) $E(Y) = \{\{x,y\} \in E \mid x \in Y,\ y \in Y\}$.

The nodes x and y of G are adjacent if $\{x,y\} \in E$. For $Y \subset X$, the adjacent set of Y, denoted by $\mathrm{Adj}(Y)$, is

(2.2.2) $\mathrm{Adj}(Y) = \{x \in X \backslash Y \mid \exists y \in Y \ni \{x,y\} \in E\}$.

When Y is a single node $\{y\}$, we write $\mathrm{Adj}(y)$ rather than $\mathrm{Adj}(\{y\})$. The degree of the node x in G, denoted by $\deg(x)$, is the number $|\mathrm{Adj}(x)|$, where $|S|$ denotes the cardinality of the set S. A graph is complete if every pair of nodes is adjacent; a subgraph G' of G is a clique in G if G' is complete.

For distinct nodes x and y in G, a path from x to y of length ℓ is an ordered set of distinct nodes $\{v_1,v_2,\ldots,v_{\ell+1}\}$ where $x = v_1$ and $y = v_{\ell+1}$, such that $\{v_i,v_{i+1}\} \in E$, $1 \le i \le \ell$. (Since we do not allow multiple edges between nodes, we could equivalently define a path as an ordered set of edges $\{v_1,v_2\},\{v_2,v_3\},\ldots\{v_\ell,v_{\ell+1}\}$. Thus, there is no confusion in referring to the "edges in a path", even though a path consists of nodes according to our definition.) A cycle is a path which begins and ends at the same node. A graph G is connected if for every pair of distinct vertices $x,y \in X$, there is at least one path from x to y; otherwise, G is disconnected, and consists of two or more connected components.

The distance $d(x,y)$ between two nodes x and y in the connected graph G is the length of the shortest path connecting them. The eccentricity of a node x is the quantity

(2.2.3) $e(x) = \max\{d(x,y) \mid y \in X\}$.

The diameter of G is then defined as

(2.2.4) $\delta(G) = \max\{e(x) \mid x \in X\}$.

A node $x \in X$ is a peripheral node of G if $e(x) = \delta(G)$.

The set $Y \subset X$ is a separator of the connected graph G if the section graph $G(X \backslash Y)$ consists of two or more components; Y is a minimal separator if no proper subset of Y is a separator of G.

The notion of a level structure [3] is an important construct in several of the algorithms we discuss. A level structure of a graph $G = (X,E)$ is a partition

$$(2.2.5) \quad \mathcal{L} = \{L_0, L_1, \ldots, L_\ell\}$$

of the node set X such that

$$(2.2.6) \quad \begin{array}{l} Adj(L_i) \subseteq L_{i-1} \cup L_{i+1}, \; 0 < i < \ell \\ Adj(L_0) \subseteq L_1, \; Adj(L_\ell) \subseteq L_{\ell-1} \end{array}$$

Obviously each L_i, $0 < i < \ell$, is a separator of G, although not necessarily minimal. The number $\ell = \ell(\mathcal{L})$ is called the length of the level structure, and the width $w = w(\mathcal{L})$ of the level structure is defined by

$$(2.2.7) \quad w(\mathcal{L}) = \max\{|L_i| \,|\, L_i \in \mathcal{L}\}.$$

For $x \in X$, the rooted level structure at x is

$$\mathcal{L}(x) = \{L_0(x), L_1(x), \ldots, L_\ell(x)\},$$

where $L_0(x) = \{x\}$ and $L_i(x) = Adj(\bigcup_{k=0}^{i-1} L_k(x))$.

A tree $T = (X,E)$ is a connected graph with no cycles. It is easy to verify that for a tree $T = (X,E)$, $|X| = |E|+1$, and every pair of distinct nodes of T are connected by exactly one path [4]. A rooted tree is an ordered pair (R,T), where R is a node of $T = (X,E)$. The node R is called the root. Since every pair of nodes in T is connected by exactly one path, the path from R to any node $x \in X$ is unique. If this path passes through y, x is said to be a descendant of y, and y is an ancestor of x. If $\{x,y\} \in E$, then y is the father of x, and x is a son of y. The section graph T(Y), where Y consists of a node and all its descendants, is called a subtree of T. Note that the ancestor-descendant and father-son relationships are only defined for rooted trees.

Given a $G = (X,E)$, let P be a partition of the node set X:

$$P = \{Y_1, Y_2, \ldots, Y_p\}.$$

That is, $X = \bigcup_{i=1}^{p} Y_i$ and $Y_i \cap Y_j = \phi$, $i \neq j$. The quotient graph of G with respect to P, denoted by G/P, is the graph

$$(2.2.8) \quad G/P = \{P, \mathcal{E}\},$$

where $\{Y_i, Y_j\} \in \mathcal{E}$ if and only if $Adj(Y_i) \cap Y_j \neq \phi$. When G/P is a tree, we call P a tree partitioning and G/P a quotient tree. A tree partitioning $P = \{Y_1, Y_2, \ldots, Y_p\}$ is maximal if no finer partitioning $Q = \{Z_1, Z_2, \ldots, Z_q\}$, (where $q > p$ and for each i and some j, $Z_i \subseteq Y_j$) yields a quotient tree.

For a graph $G = (X,E)$ with $|X| = N$, an <u>ordering</u> (numbering, labelling) of G is a bijective mapping $\alpha:\{1,2,\ldots,N\} \rightarrow X$. We denote the labelled graph and node set by G^{α} and X^{α} respectively. An ordering α is <u>compatible</u> with a partitioning P if for each $Y \in P$, α numbers the nodes of Y consecutively. An ordering of G which is compatible with a partitioning P <u>induces</u> an ordering on the quotient graph G/P. Conversely, an ordering of G/P in general induces a <u>class</u> of orderings of G which are compatible with P.

Let A be an N by N symmetric matrix. The labelled undirected graph of A is denoted by $G(A) = (X(A),E(A))$ and is one for which $X(A) = \{x_1,x_2,\ldots,x_N\}$, and $\{x_i,x_j\} \in E(A) \iff A_{ij} \neq 0$, $i \neq j$.[†] The unlabelled graph of A is simply $G(A)$ with its labels removed. For any N by N permutation matrix $P \neq I$, the unlabelled graphs of A and PAP^T are the same, but the associated labellings are different. Thus, finding a "good" ordering for A can be viewed as finding a good labelling for its graph. As suggested by the remarks of section 1.3, the goodness of an ordering is a complicated function of how we intend to store and process the matrix and its factors, the type of problem we have, the computer charging algorithm, etc.

2.3 <u>Computer Representation of Graphs</u>

In subsequent sections we will be describing algorithms for finding orderings for graphs, and also reporting computer execution times for those algorithms. The performance of such algorithms is sometimes quite sensitive to the way the graphs are represented, so in this section we provide a brief description of some useful data structures for graphs. As usual, we denote the graph by $G = (X,E)$, with $|X| = N$.

An <u>adjacency list</u> for $x \in X$ is a list containing all the nodes in $Adj(x)$. An <u>adjacency structure</u> for G is simply the set of adjacency lists for all $x \in X$. Such a structure can be implemented quite simply and economically by storing the adjacency lists sequentially in a one-dimensional array ALIST along with an index array XLIST length N containing pointers to the positions of each adjacency list in ALIST. An example is shown in Figure 2.3.1.

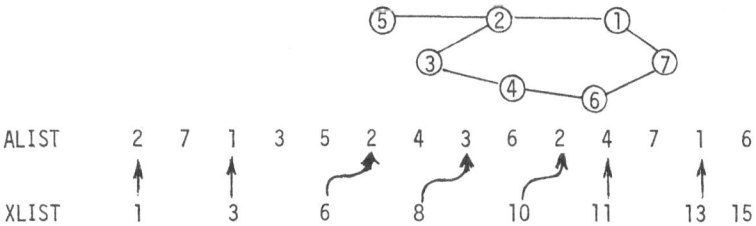

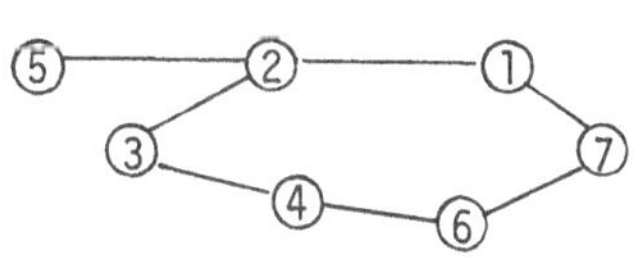

ALIST	2	7	1	3	5	2	4	3	6	2	4	7	1	6
XLIST	1		3		6		8		10		11		13	15

Figure 2.3.1 Example of an adjacency structure

---

† There should be no confusion between the designation of the graph of a matrix $G(A)$ and the section graph $G(A)$, since the argument in the first case must be a matrix, and in the second case it must be a set of nodes of the graph G.

It is often convenient for coding purposes to have an extra entry in XLIST such that XLIST(N+1) points to the next available storage location in ALIST, as shown in Figure 2.3.1. Clearly the total storage requirement for this storage scheme is then $|X|+2|E|+1$.

A second common storage scheme is a simple connection table, having N rows and m columns, where $m = \max\{\deg(x) | x \in X\}$. The adjacency list for node i is stored in row i. This storage scheme may be quite inefficient if a substantial number of the nodes have degrees less than m.

Any variant of the above two schemes possesses an important disadvantage. Unless the degrees of the nodes are known a priori, it is difficult to construct the storage scheme when the graph is provided as a list of edges. In this case a link field is required, as described in [28]. Figure 2.3.2 is an example of such a scheme for the graph of Figure 2.3.1. The pointer HEAD(i) starts the adjacency list for node i, with NBRS containing the neighbor of node i and LINK containing the pointer to the location of the next neighbor of node i. A negative link of -i indicates the end of the adjacency list for node i. The storage requirement for this graph representation is $|X|+4|E|$.

	HEAD
1	10
2	6
3	1
4	12
5	8
6	5
7	7

	NBRS	LINK
1	2	14
2	3	-4
3	2	-1
4	1	-7
5	4	9
6	1	13
7	6	4
8	2	-5
9	7	-6
10	7	3
11	3	-2
12	6	2
13	5	11
14	4	-3

Figure 2.3.2 Adjacency linked lists for the graph of Figure 2.3.1

The implementations of the algorithms we describe in later sections all used the connection table representation. For the problems with $t > 2$, this representation is considerably less efficient than the adjacency structure in terms of storage, but independent experiments indicate that the execution times of the algorithms are virtually identical for the two representations. Moreover, the actual computer implementations of the ordering algorithms we describe in subsequent sections all require roughly the same amount of storage.

2.4 Finding Pseudo-Peripheral Nodes of a Graph

The effectiveness of many ordering algorithms, including the ones we describe in subsequent sections, depends quite critically on the proper choice of a "starting node". A substantial amount of experience has led researchers to advocate the use of peripheral nodes as starting nodes for some algorithms. Unfortunately, no efficient algorithm is known which guarantees to find peripheral nodes of a general graph. However, a recent algorithm due to Gibbs, Poole, and Stockmeyer [21] appears to be extremely effective in finding nodes of finite element graphs whose eccentricity is close to the diameter of the graph. Our objective in this section is to describe this important algorithm.

Following [21], a pseudo-peripheral node is a node x that satisfies the condition that for any $y \in X \ni d(x,y) = e(x) \Rightarrow e(y) = e(x)$. Obviously, any peripheral node is a pseudo-peripheral node. The algorithm of Gibbs, et al., which is described below, is based on the observation that if y is a node in the last level of a level structure rooted at x, then $e(x) \leq e(y)$.

Step 1: Find a node r of minimum degree.

Step 2: Generate the rooted level structure at r:
$\mathcal{L}(r) = \{L_0(r), L_1(r), \ldots, L_{\ell(r)}(r)\}$.

Step 3: Sort $L_{\ell(r)}(r)$ in order of increasing degree.

Step 4: For each $x \in L_{\ell(r)}(r)$ in order of increasing degree, generate the level structure $\mathcal{L}(x)$ at x. If $\ell(x) > \ell(r)$, put $r \leftarrow x$ and go to step 3.

Step 5: r is a pseudo-peripheral node.

We now provide a simple example describing the operation of the algorithm. Referring to the simple graph of Figure 2.4.1, the algorithm proceeds as follows.

Step 1: Node 8 chosen as starting node, since it is one of minimum degree.

Step 2: Level structure rooted at 8 yields a last level containing nodes {1,14,15,16}. The length of the level structure is 4.

Step 3: Nodes of the last level are sorted by degree, yielding the order {16,1,14,15}.

Step 4: Level structure rooted at 16 has a length of 7, so step 3 is begun with the last level, which is {1}.

Step 3: The set {1} is already sorted.

Step 4: The level structure rooted at 1 has length 7, so 1 is a pseudo-peripheral node.

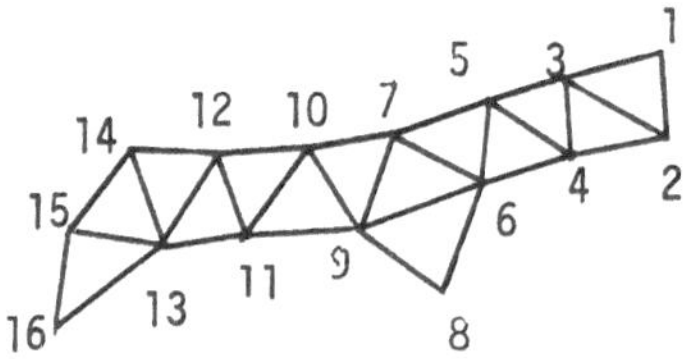

Figure 2.4.1 A simple graph to illustrate the use of the algorithm of Gibbs, et al.

In their actual computer implementation of the algorithm [6], the execution time is significantly reduced by rejecting wide level structures as soon as they are detected. (In their application, they are interested in finding narrow level structures, rather than pseudo-peripheral nodes.) In other words, for many of the $x \in L_{\ell(r)}(r)$, only part of the level structure $\mathcal{L}(x)$ will be generated.

In all our applications we found that the application of this "short-circuit" technique did not substantially affect the quality of the ordering produced, so our implementation also includes the device. Of course, the modified algorithm may no longer provide a pseudo-peripheral node.

We also have incorporated another modification into the algorithm which in some situations can substantially reduce the execution time, and again, for our application, did not significantly affect the outcome. For some problems, the set $L_{\ell(r)}(r)$ can be quite large so that step 4 has to be executed $|L_{\ell(r)}(r)|$ number of times. In our implementation, we modify the algorithm by picking representatives from $L_{\ell(r)}(r)$. Step 3 and step 4 are replaced as follows.

<u>Step 3'</u>: Find all the connected components in $L_{\ell(r)}(r)$.

<u>Step 4'</u>: For each component C in $L_{\ell(r)}(r)$, find a node x of minimum degree in C and generate its rooted level structure $\mathcal{L}(x)$. If $\ell(x) > \ell(r)$, put $r \leftarrow x$ and go to step 3'.

In the sequel we will refer to nodes found using this algorithm as "pseudo-peripheral" nodes. However, it should be understood that they may not satisfy the definition given above because of the modifications to the original algorithm.

3. Profile (Envelope) Methods

3.1 Preliminaries

A very common and widely used ordering strategy for finite element equations is one for which the coefficient matrix A has a small <u>bandwidth</u> or <u>profile</u>. For row i of the N by N symmetric matrix A define

(3.1.1) $f_i(A) = \min\{j \mid A_{ij} \neq 0\}$,

and

$$(3.1.2)\qquad \beta_i(A) = i - f_i(A).$$

Obviously $f_i(A)$ is just the column subscript of the first nonzero in row i of A. The bandwidth $\beta(A)$ is then defined as

$$(3.1.3)\qquad \beta(A) = \max\{\beta_i(A),\ 1 \le i \le N\}.$$

The quantity $\beta_i(A)$ is referred to as the i-th bandwidth of A. We define Band(A) by

$$(3.1.4)\qquad \mathrm{Band}(A) = \{\{i,j\} \mid 0 < i-j \le \beta(A)\}.$$

This set of subscripts of A corresponds to those components of A within $\beta(A)$ of the diagonal.

Typical finite element applications very often involve irregular meshes having, for example, appendages and/or holes. It is also becoming increasingly common for finite element applications to involve elaborate elements having numerous side and interior nodes. In these situations there will generally be considerable variation in the individual row bandwidth, and it is natural to exploit this variation in both computation and storage. In this connection we make the following definitions.

The envelope of A is defined by

$$(3.1.5)\qquad \mathrm{Env}(A) = \{\{i,j\} \mid 0 < i-j \le \beta_i(A)\},$$

and the profile of A is simply the number

$$(3.1.6)\qquad |\mathrm{Env}(A)| = \sum_{i=1}^{N} \beta_i(A).$$

It is easy to verify that all fill suffered during the application of symmetric Gaussian elimination to A is confined to Env(A), and therefore to Band(A) since $\mathrm{Env}(A) \subseteq \mathrm{Band}(A)$.

Implicit in the use of these orderings is the assumption that A, for purposes of storage and/or computation, is to be subsequently regarded as having a potentially dense band or envelope. It is well-known that if we are prepared to exploit all zeros, then orderings which minimize $\beta(A)$ or $|\mathrm{Env}(A)|$ may be very far from optimal in the least fill or least arithmetic sense. However, they often represent a reasonable compromise between optimality in the above sense and convenient programming and storage management. Minimum or near-minimum fill orderings characteristically lead to triangular factors whose nonzero components are scattered throughout the lower triangle, so a relatively sophisticated data structure and program is required to exploit the zeros. By comparison, implementations which utilize band or envelope orderings are relatively simple. In addition, algorithms for finding small band/envelope orderings are much more widely available and more reliable than algorithms for finding near minimum fill (or arithmetic) orderings. Finally, for large problems where more sophisticated techniques are warranted, these simple band or envelope schemes are often used in a subsidiary fashion, as we shall see in sections 5 and 6.

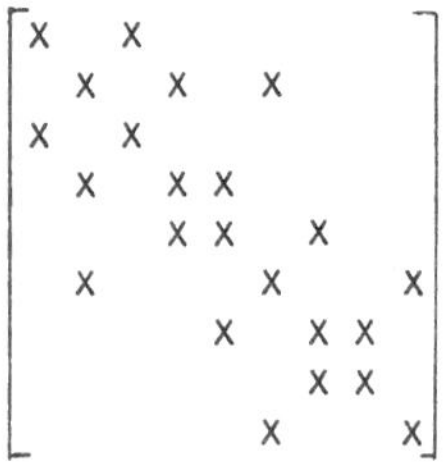

i	$\beta_i(A)$	f_i
1	0	1
2	0	2
3	2	1
4	2	2
5	1	4
6	4	2
7	2	5
8	1	7
9	3	6

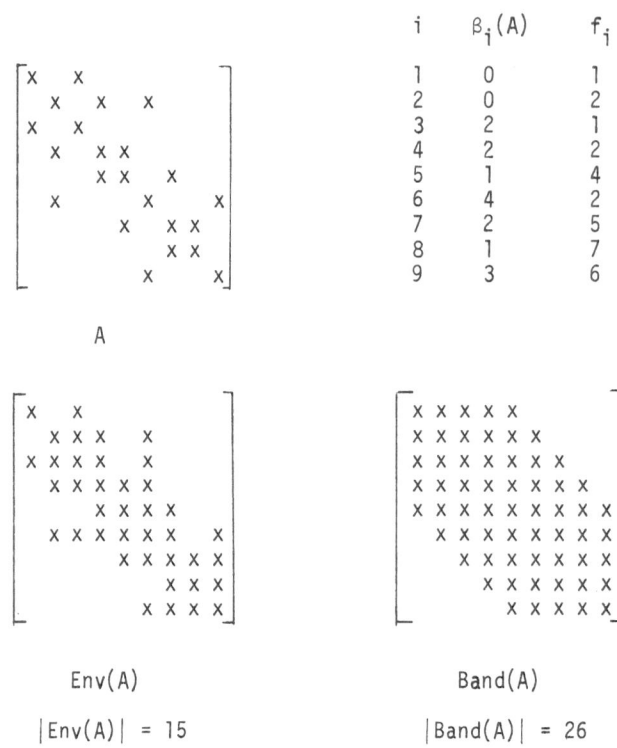

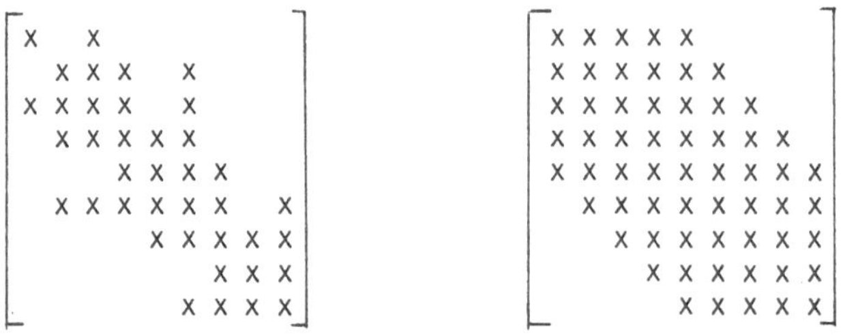

Figure 3.1.1 An example showing Env(A), Band(A), and the functions β and f.

3.2 Storage Schemes

A common method for storing a symmetric band matrix A with bandwidth $\beta(A)$ is the so-called diagonal storage scheme. The essential idea is to store the non-null diagonals of the lower triangle including the diagonal, as the columns of an N by $\beta(A)+1$ rectangular array. See the example in Figure 3.2.1. However, as we mentioned in section 3.1, in many situations there will be considerable variation in the $\beta_i(A)$, and the diagonal storage scheme may be consequently quite wasteful. This variation can be exploited by using a scheme proposed by Jennings [23]. For each row in A, all the entries from the first nonzero to the diagonal are stored in contiguous locations in a one-dimensional array S. An additional N pointers are required to record the positions of the diagonal components of A in S. Since the fill is confined to Env(A), the factorization can be carried out "in place". Figure 3.2.1 contains an illustration of this storage scheme. This storage scheme is very appropriate in many circumstances since it can often be shown that $Env(L+L^T)$ has no zeros [16].

3.3 The (Reverse) Cuthill-McKee Algorithm

The Cuthill-McKee algorithm [8] is an ordering algorithm designed to find an ordering having a small bandwidth. A large amount of experience with finite element problems has established it as the industry standard. In [12], the author provided some experiments demonstrating that by reversing the ordering provided by the Cuthill-McKee algorithm, an ordering having a much smaller profile could often be obtained. The bandwidth of course remains unchanged. We refer to this ordering algorithm as the Reverse Cuthill-McKee algorithm (RCM). More recently, it has been shown that this reversal can never increase the profile [24].

$$\begin{bmatrix} A_{11} & & & & & & \\ A_{21} & A_{22} & & & & & \text{Symmetric} \\ & & A_{33} & & & & \\ A_{41} & & & A_{44} & & & \\ & & A_{53} & A_{54} & A_{55} & & \\ & & A_{63} & & & A_{66} & \\ & & & & A_{75} & A_{76} & A_{77} \end{bmatrix} \qquad \begin{bmatrix} & & & A_{11} \\ & & A_{21} & A_{22} \\ & 0 & 0 & A_{33} \\ A_{41} & 0 & 0 & A_{44} \\ 0 & A_{53} & A_{54} & A_{55} \\ A_{63} & 0 & 0 & A_{66} \\ 0 & A_{75} & A_{76} & A_{77} \end{bmatrix}$$

A

Diagonal storage scheme

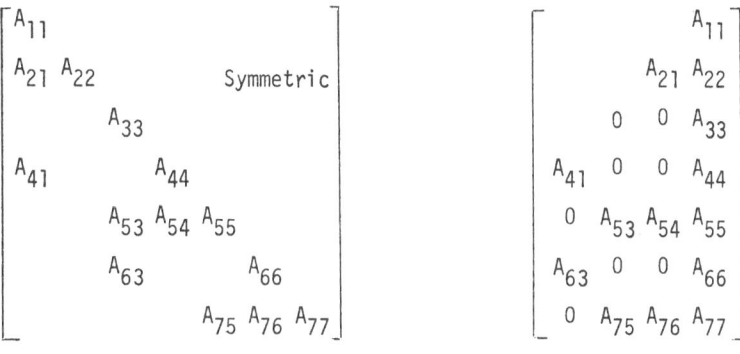

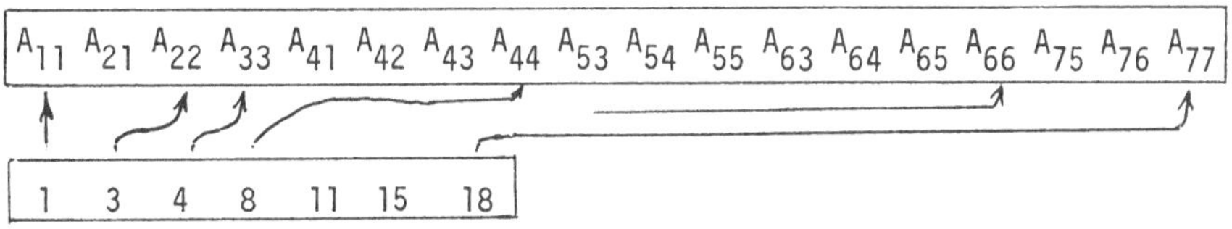

Jennings' storage scheme

Figure 3.2.1 An example of the diagonal storage scheme and Jennings' storage scheme

However, the algorithm requires a starting node, and this search for a starting node is an expensive task in current industrial implementations of the algorithm [21]. The very efficient algorithm due to Gibbs, et al. (described in section 2.5) is a natural candidate for this task, and in [17] this combination of algorithms was compared with a variety of others and for our class of problems was found in balance to be the best in terms of execution time and the size of profile obtained. For a connected graph $G = (X,E)$, the overall scheme can be concisely described as follows:

<u>Step 1</u>: Determine a pseudo-peripheral node r and assign it to x_1.

<u>Step 2</u>: For $i = 1,2,\ldots,N$, find all the unnumbered neighbors of the node x_i and number them in increasing order of degree.

<u>Step 3</u>: The RCM ordering is given by: $x_N, x_{N-1}, \ldots, x_1$.

3.4 Numerical Experiments

We begin this section by providing some numerical results to show the rather substantial savings in storage that can be achieved by exploiting the variation in the bandwidth; that is, by using Jennings' storage scheme rather than the diagonal storage scheme, both of which were described in the previous section. In Table 3.4.1 the operations column (θ_B) for the diagonal storage scheme assumes that no zeros within the band are exploited, and serves to show that even if the diagonal storage scheme is used, it is still very worthwhile to exploit zeros within the band. Since a check for a zero "multiplier" is relatively inexpensive in terms of execution time, most programs only perform about θ_E operations, even if the entire band is stored. Note that the advantage of the envelope method over the standard band scheme is much greater for the more elaborate (t = 2) elements and for the domains with appendages.

TABLE 3.4.1

Comparison of band and envelope schemes with respect to storage and operations

		ENVELOPE		BAND	
Example	Bandwidth	Storage	Operations (θ_E)	Storage	Operations (θ_B)
a1	33	26,645	344,608	37,026	632,709
a2	65	24,764	334,114	63,426	2,023,818
b1	35	31,040	487,992	36,324	654,243
b2	69	28,988	422,357	70,630	2,387,331
c1	31	27,043	332,412	37,760	610,007
c2	58	20,810	241,666	55,755	1,599,940
d1	27	23,062	195,063	38,556	549,666
d2	43	11,804	85,875	35,420	765,944
e1	27	22,753	301,788	26,208	371,061
e2	58	25,781	356,872	55,224	1,584,018
f1	21	29,661	300,226	31,680	358,911
f2	44	23,355	232,144	51,840	1,158,901

It is interesting to investigate the time and space demands that this combination of algorithms is likely to have for our class of problems. In order to obtain some insight into this question, several of the test problems were solved with increasing mesh subdivision factor s. The behavior was very similar for each problem, as typified by the results displayed in Table 3.4.2.

Several points about the columns of Table 3.4.2 are noteworthy. First, the execution time of the ordering algorithm appears to be linear in N. Second, the operation count and factorization times increase quadratically with N. As N increases, an increasing proportion of the execution is confined to the innermost

TABLE 3.4.2 Performance statistics of the RCM ordering algorithm coupled with the linear equations solver ENVSLV which utilizes Jennings' storage scheme. The statistics are for the Graded L domain with s=4(1)12 and t=1

		Order		Factorization				Solution				Storage	
S	N	Time	Time/N	Time	Time/N^2	Operations	Time/Operations	Time	Time/$N^{3/2}$	Operations	Time/Operations	Total	Total/$N^{3/2}$
4	265	.08	3.02(-4)	.42	5.98(-6)	37,426	11.22(-6)	.08	1.85(-5)	8,412	9.51(-6)	4,474	1.037
5	406	.11	2.71(-4)	.89	5.40(-6)	84,381	10.55(-6)	.14	1.71(-5)	15,710	8.91(-6)	8,264	1.010
6	577	.16	2.77(-4)	1.64	4.92(-6)	165,659	9.90(-6)	.22	1.59(-5)	26,334	8.35(-6)	13,747	.992
7	779	.21	2.69(-4)	2.82	4.66(-6)	294,892	9.56(-6)	.33	1.52(-5)	40,908	8.07(-6)	21,235	.979
8	1009	.28	2.77(-4)	4.41	4.33(-6)	487,992	9.04(-6)	.47	1.47(-5)	60,056	7.83(-6)	31,040	.968
9	1270	.34	2.68(-4)	6.75	4.18(-6)	763,151	8.85(-6)	.66	1.46(-5)	84,402	7.82(-6)	43,474	.961
10	1561	.42	2.69(-4)	9.91	4.06(-6)	1,140,841	8.69(-6)	.88	1.43(-5)	114,570	7.68(-6)	58,849	.954
11	1882	.51	2.71(-4)	14.06	3.97(-6)	1,643,814	8.55(-6)	1.16	1.42(-5)	147,420	7.87(-6)	77,477	.949
12	2233	.60	2.69(-4)	19.41	3.89(-6)	2,297,102	8.45(-6)	1.49	1.41(-5)	190,402	7.83(-6)	99,670	.945

loop, which explains why the time-per-operation tends to drop slowly with N. (A substantial component of the execution overhead grows only as $N^{3/2}$.) Similar remarks apply to the results for the solution of the equations, given the factorization.

We report only the total storage in Table 3.4.2 because the overhead storage is essentially negligible - only N pointers. The total storage requirement appears to increase as $N^{3/2}$.

Although the actual constants of proportionality varied somewhat over the different domains, the asymptotic behavior of the ordering and solving programs was the same. Note that this behavior corresponds to what is known to be the case for model n by n ($N = n^2$) grid problems [13], when the row by row ("natural") ordering is used.

4. Storage and Solution of Partitioned Matrix Problems

4.1 Preliminaries

In section 2 we defined a partitioning of a graph, and orderings which were compatible with such a partitioning. Given the correspondence that we have established between graphs and matrices in section 2, a partitioning P and a compatible ordering α of G, where G is the graph of some matrix, specifies a particular ordering and partitioning of the matrix. Partitionings of matrices obtained this way are obviously symmetric in the sense that the row and column partitionings are identical. The objective of this section is to describe some important advantages of storing and processing matrices in partitioned form. This will provide the motivation for developing algorithms for finding certain orderings and partitionings of matrices (graphs), which are the subjects of sections 5 and 6.

4.2 Basic Observations

Suppose the sparse matrix equation Ax = b is partitioned as shown in equation (4.2.1) below.

$$(4.2.1) \quad \begin{pmatrix} A_{11} & A_{12} \\ A_{12}^T & A_{22} \end{pmatrix} \begin{pmatrix} x_1 \\ x_2 \end{pmatrix} = \begin{pmatrix} b_1 \\ b_2 \end{pmatrix} .$$

The Cholesky factor L of A, correspondingly partitioned, is

$$(4.2.2) \quad L = \begin{pmatrix} L_{11} & 0 \\ W_{12}^T & L_{22} \end{pmatrix},$$

where $L_{11}L_{11}^T = A_{11}$, $W_{12} = L_{11}^{-1}A_{12}$, and

$$(4.2.3) \quad L_{22}L_{22}^T = \tilde{A}_{22} = A_{22}-A_{12}^TA_{11}^{-1}A_{12} = A_{22}-\hat{A}_{22}$$

Here and elsewhere in this paper it is understood that inverses are not computed explicitly; instead, the appropriate triangular systems are solved.

Since sparse matrices normally suffer fill during their factorization, the matrix W_{12} may be much less sparse than A_{12}. When this is true, the following observations become relevant [14].

Observation 1:

There are two distinctly different ways to compute the product $A_{12}^T A_{11}^{-1} A_{12}$. We could compute $W_{12} = L_{11}^{-1} A_{12}$, and then multiply W_{12}^T by W_{12}. This is the way the computation is done if the ordinary standard Cholesky algorithm is used. However, it may require substantially fewer arithmetic operations to compute $\hat{A}_{22}$ as $A_{12}^T(L_{11}^{-T}(L_{11}^{-1}A_{12}))$. It is easy to see this with an example where W_{12} turns out to be full, but A_{12} and L_{11} are themselves very sparse. Note that the symmetry of the matrix $\hat{A}_{22}$ can still be exploited.

This asymmetric version of the computation has an advantage in terms of storage management as well. In some situations, discussed under observation 2 below, we do not wish to retain the off-diagonal block(s) of the factorization. In this case, the asymmetric computational scheme allows us to compute one column of $\hat{A}_{22}$ at a time, substract it from A_{22}, and then discard it. Thus, we need only one auxiliary vector. On the other hand, the symmetric scheme appears to require the temporary storage of all of W_{12} in order to compute $W_{12}^T W_{12}$, even though we do not intend to retain W_{12} for subsequent computation.

Observation 2:

In order to solve (4.2.1), given the factorization of A, it is necessary to multiply W_{12} and W_{12}^T by certain vectors. However, given the simple definition $W_{12} = L_{11}^{-1} A_{12}$, multiplication by W_{12} can be carried out implicitly by multiplying by A_{12} and then solving the appropriate triangular system. This technique not only can save considerable storage, but sometimes can also reduce the amount of arithmetic required.

As suggested by George [14, page 586], these ideas can be generalized in a natural way. Obviously the matrix A_{11} could itself be partitioned into a 2 by 2 block matrix, and the ideas recursively applied. In general, such a scheme involves a high degree of recursion, but in the special case that the quotient graph corresponding to the partitioning is an appropriately ordered tree, no recursion at all is necessary. We explore these ideas in section 5.

4.3 Two Storage Schemes for Partitioned Matrices

In this section we describe two storage schemes for partitioned matrices which are particularly appropriate for the orderings and partitionings we describe in sections 5 and 6. For illustrative purposes we assume A is partitioned into p^2 submatrices A_{rs}, $1 \le r,s \le p$, and let L_{rs} be the corresponding submatrices of L, where $A = LL^T$. Define the matrices

$$(4.3.1) \qquad B_k = \begin{bmatrix} A_{1k} \\ A_{2k} \\ \vdots \\ A_{k-1,k} \end{bmatrix} \qquad 2 \le k \le p.$$

Thus, A can be viewed as follows, where p is chosen to be 5.

$$(4.3.2) \qquad \left[\begin{array}{c|c|c|c|c} A_{11} & B_2 & B_3 & B_4 & B_5 \\ A_{12}^T & A_{22} & & & \\ A_{13}^T & A_{23}^T & A_{33} & & \\ A_{14}^T & A_{24}^T & A_{34}^T & A_{44} & \\ A_{15}^T & A_{25}^T & A_{35}^T & A_{45}^T & A_{55} \end{array}\right]$$

As we mentioned earlier, and will see in more detail later, one computational scheme requires that we store the diagonal blocks L_{ii}, $1 \le i \le p$ of the triangular factor L, and the off-diagonal blocks of the original matrix A. The storage scheme we advocate, called the implicit storage scheme, is illustrated by the example of Figure 4.3.1, which is taken from [17]. The diagonal blocks of L are viewed as forming a single matrix and stored row by row in a single array ν, as proposed by Jennings and described in section 3. An additional vector δ of length N records the positions of the diagonal components, and another vector τ of length p records the beginning of each diagonal block in δ. For programming convenience we set $\tau_{p+1} = N+1$.

The nonzero components of B_i, $1 < i \le p$ are stored in a single one dimensional array ξ, column by column. A parallel integer array ρ stores the row subscripts of those numbers in ξ. A vector γ of length N records the positions in ξ where each column starts. For programming convenience we set $\gamma_{N+1} = |\rho|+1$, where $|\rho|$ denotes the number of components in ρ. Note that $\Delta\gamma_i = \gamma_{i+1}-\gamma_i = 0$ implies that the corresponding column of B_r is null, where $\tau_r \le i < \tau_{r+1}$. Also note that $\gamma_{\tau_{r+1}}-\gamma_{\tau_r} = 0$ implies that the entire matrix B_r is null. Obviously, the storage required for the vectors δ, τ, ρ and γ must be regarded as overhead storage, since such storage is not used to store actual matrix components.

We now turn to a second storage scheme, which for reasons which will be immediately apparent, we call the block-within-block-column scheme. It is used in connection with the dissection ordering strategy of section 6. This scheme stores the diagonal blocks in exactly the same way as in the implicit storage scheme just described, using the vectors ν, δ and τ. Let C_k be defined by

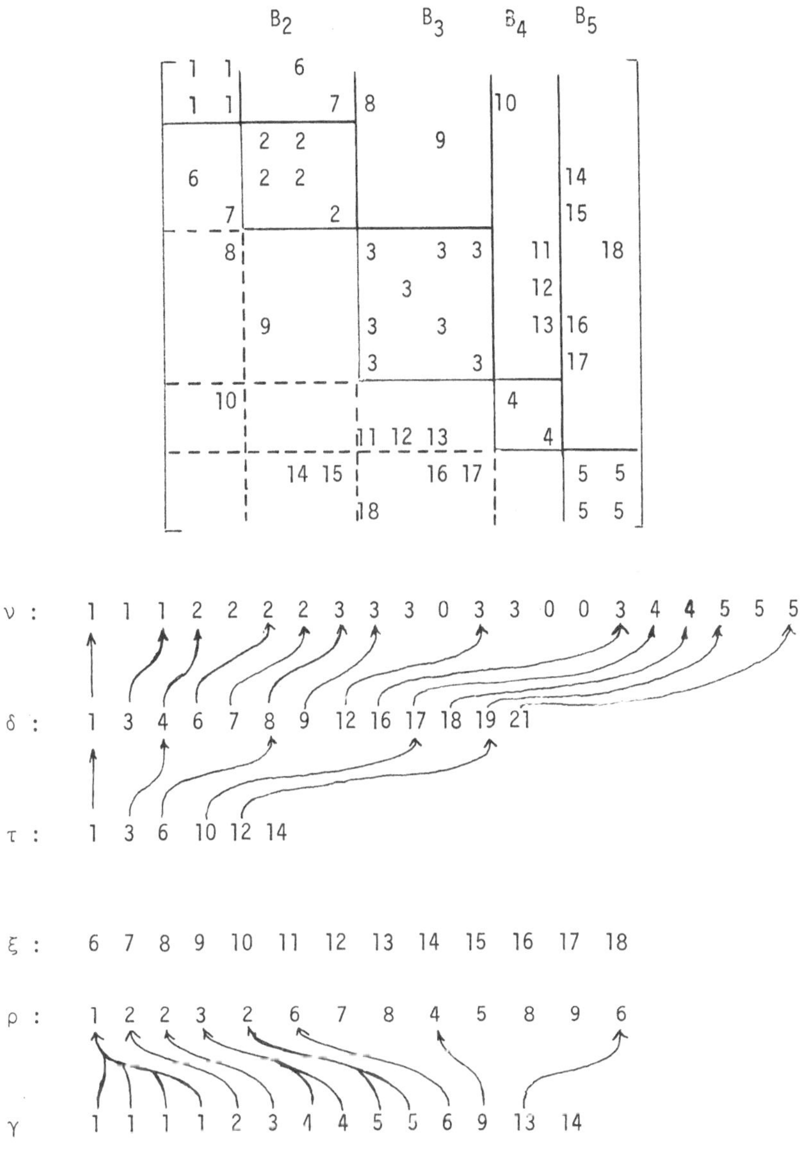

Figure 4.3.1 Example showing the arrays used in the implicit storage scheme

$$(4.3.3) \quad C_k = \begin{bmatrix} L_{k+1,k} \\ L_{k+2,k} \\ \vdots \\ L_{p,k} \end{bmatrix}, \quad 1 \le k < p$$

The matrix C_k is obviously the part of the k-th block-column which is below the diagonal block $L_{k,k}$ in the partitioned matrix L. Now in the application we have in mind, the nonzeros in each block column are themselves typically arranged in blocks; each row segment of L in a block-column is either empty or full, and non-null row segments cluster together in consecutive positions. The storage scheme that is naturally suggested for the off-diagonal blocks is illustrated in Figure 4.3.2.

The vector ξ is used to store the μ off-diagonal blocks (within the block-columns) of L, stored column by column in consecutive locations. The vector γ points to the origins of the blocks in ξ, and the parallel vector ρ indicates the row number (in L) of the first row of each block. Finally, in order to specify which block column of L a particular block in ξ resides, we have a vector σ of length p which points into γ; specifically, the origins in ξ of the blocks of block-column i of L are given by γ_ℓ, $\sigma_i \le \ell < \sigma_{i+1}$. Again for programming convenience, we set $\gamma_{\mu+1} = |\xi|+1$ and $\sigma_p = \mu+1$. Note that for $\sigma_i \le \ell < \sigma_{i+1}$, the number of rows in the block stored beginning at ξ_{γ_ℓ} is simply $(\gamma_{\ell+1}-\gamma_\ell)/(\tau_{i+1}-\tau_i)$. Note that the storage needed for the vectors δ, τ, γ, ρ and σ is overhead storage.

5. On Trees and Tree Partitionings for Finite Element Problems

5.1 Introduction

In this section we consider matrices whose graphs are trees. Such matrices have the important property that they can be ordered so that no fill occurs when Gaussian elimination is performed. Unfortunately, the matrices which arise in typical finite element applications do not have graphs which are trees. The objective of this section is to describe a scheme for finding a partitioning P of a finite element graph G so that the quotient graph G/P is a tree. Many of the desirable features of trees can then be exploited in solving the matrix problem, using an ordering compatible with P. We also provide numerical experiments comparing methods of this section to the envelope ordering/solution scheme of section 3.

5.2 On Matrices whose Graphs are Trees

Given a tree T = (X,E), a monotone ordering α for the rooted tree (R,T) is one for which each node is numbered before its father. Obviously, the root R must be numbered last. Given an unrooted tree T, the ordering α is monotone if it is monotone for the rooted tree $(\alpha(|X|),T)$.

Lemma 5.2.1 (Parter [27]) Let A be an N by N symmetric matrix whose graph is a mono-

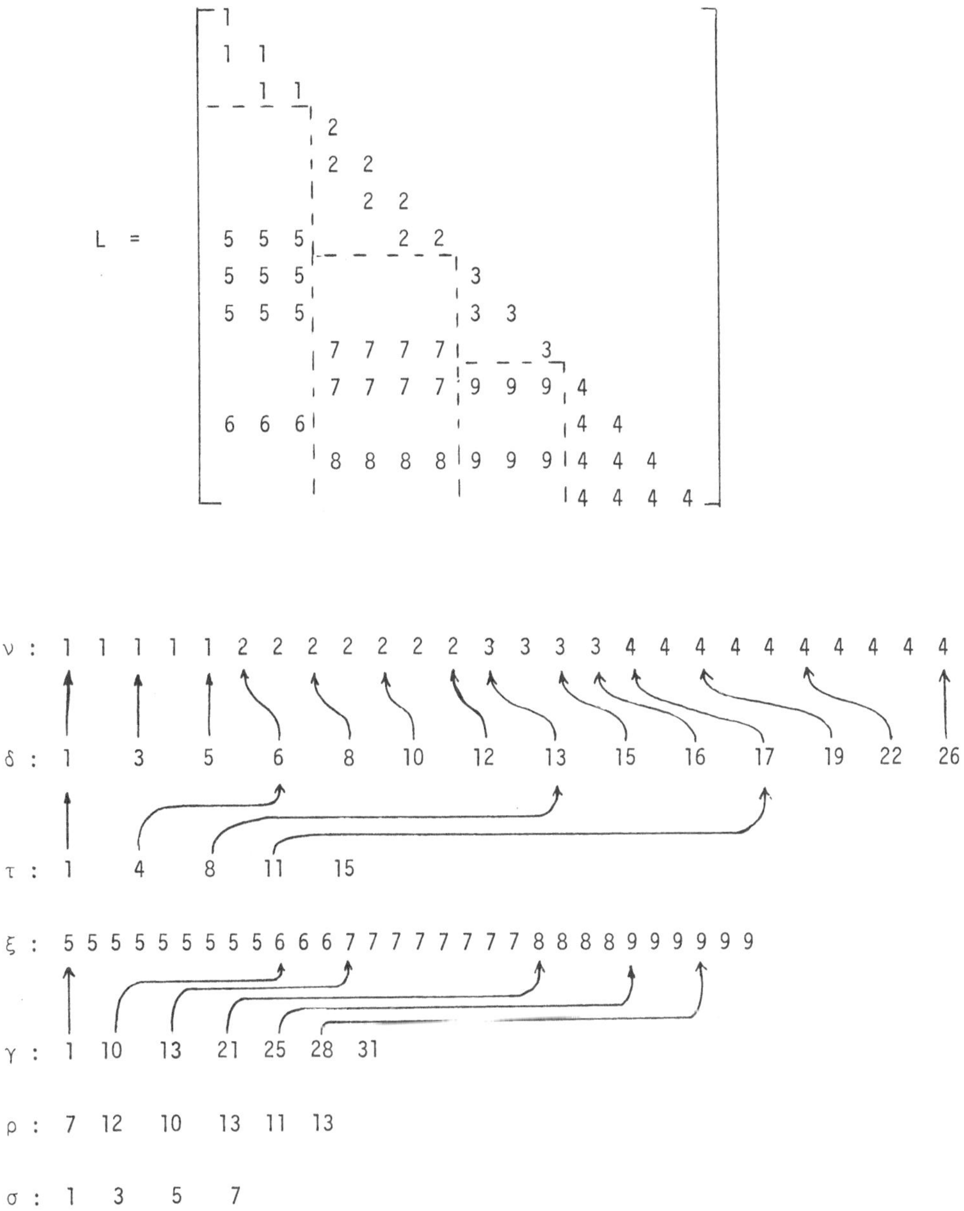

Figure 4.3.2 Block-within-block-column storage scheme

tonely ordered tree. If $A = LL^T$, where L is the Cholesky factor of A, then $A_{ij} = 0 \Rightarrow L_{ij} = 0$, $i > j$. □

In other words, as we mentioned above, matrices associated with monotonely ordered trees suffer no fill during Gaussian elimination.

Lemma 5.2.2 [17] Let A and L be as in Lemma 5.2.1. Then $L_{ij} = A_{ij}/L_{jj}$, $i > j$.

Proof The components of L are given by

$$L_{ij} = (A_{ij} - \sum_{k=1}^{j=1} L_{ik}L_{jk})/L_{jj}, \quad i > j.$$

It is sufficient to show that $\sum_{k=1}^{j-1} L_{ik}L_{jk} = 0$. Suppose for a contradiction that $L_{im}L_{jm} \neq 0$ for some m satisfying $1 \leq m \leq j-1$. By Lemma 5.2.1, this implies $A_{im}A_{jm} \neq 0$, so node x_m is connected to both x_i and x_j, $i > m$, $j > m$. Thus, the graph of A is not monotonely ordered.

Lemmas 5.2.1 and 5.2.2 extend immediately to partitioned matrices, and it is in this context that Lemma 5.2.2 has its most important implication. As in section 4.3, assume A is partitioned into p^2 submatrices A_{ij}, $1 \leq i,j \leq p$, and let L_{ij} be the corresponding submatrices of L. In addition, we assume that the quotient graph of the partitioned matrix A is a monotonely ordered tree, as illustrated in Figure 5.2.1. It is then straightforward to verify that the analog of Lemma 5.2.2 for such a partitioned matrix states that $L_{ij} = A_{ij}L_{jj}^{-T} = (L_{jj}^{-1}A_{ji})^T$ for each non-null submatrix A_{ij} in the lower triangle of A. Thus, for such partitioned matrices,the implicit storage scheme described in section 4.3 is highly appropriate,and the observations of section 4.2 are applicable.

The previous lemmas and discussion motivate the next section, which describes an algorithm for finding a tree partitioning (and compatible ordering) of a given graph G.

5.3 An Algorithm for Finding a Tree Partitioning and an Ordering

The results of section 5.2 suggest that in order to use the implicit storage scheme effectively, we should find a partitioning $P = \{Y_1, Y_2, \ldots, Y_p\}$ with as many members as possible, consistent with the requirement than G/P remains a quotient tree. In this section we provide an algorithm for finding a tree partitioning of a graph, based on rooted level structures.

Let G(A) be the graph associated with the symmetric matrix A and let $\mathcal{L}$ be a level structure. From the definition of a level structure (see section 2.2), it is clear that the quotient graph $G(A)/\mathcal{L}$ is a simple chain. Thus, if we number the nodes in each level L_i consecutively from L_0 to the last level, the levels of $\mathcal{L}$ induce a block tridiagonal partitioning on the correspondingly ordered matrix.

The algorithm we propose here can be viewed as a refinement of the levels of a rooted level structure. Let $\mathcal{L} = \{L_0, L_1, \ldots, L_\ell\}$ be a rooted level structure and let $P = \{Y_1, Y_2, \ldots, Y_p\}$ be the partitioning obtained by subdividing each L_j as follows. Letting G_j be the section graph (see section 2.2) prescribed by

$$A = \left[\begin{array}{cc|ccc|cc|ccc}
x & x & x & & x & & & & & \\
x & x & x & & & & & & & \\
\hline
x & x & x & x & & & & & & \\
 & & x & x & x & & & & x & \\
x & & & x & x & & & x & & \\
\hline
 & & & & & x & x & x & & \\
 & & & & & x & x & & & x \\
\hline
 & & & & x & x & & x & x & \\
 & & & x & & & & x & x & x \\
 & & & & & & x & & x & x
\end{array}\right]$$

$P = \{\{1,2\},\{3,4,5\},\{6,7\},\{8,9,10\}\}$

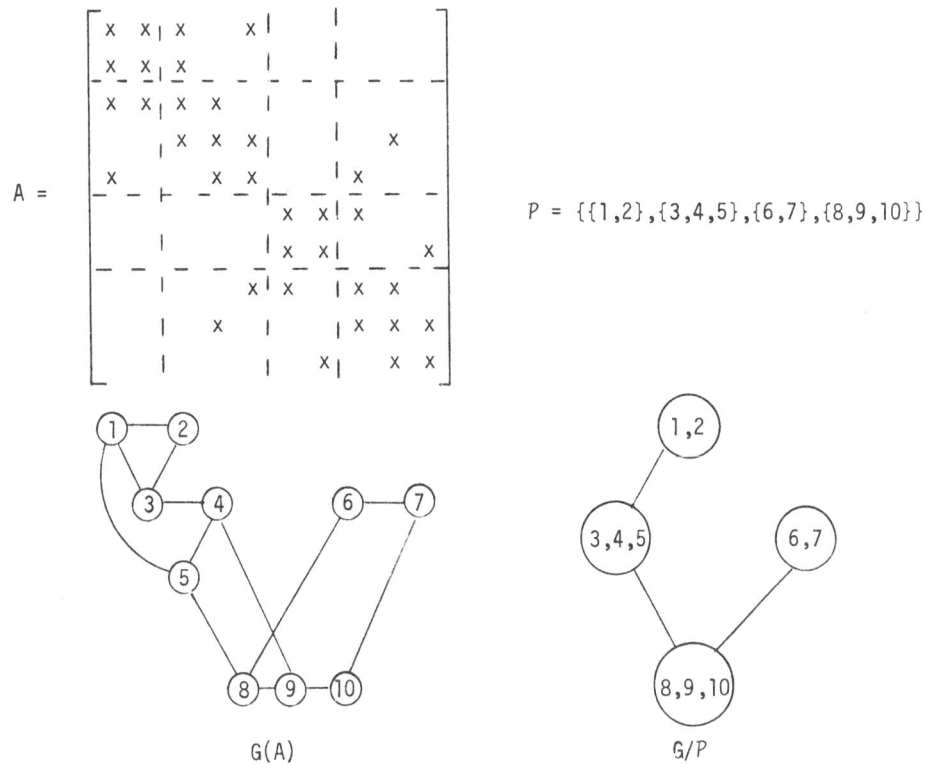

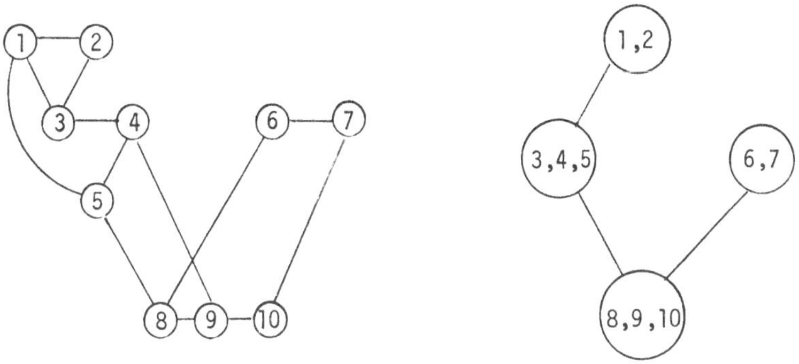

G(A) G/P

Figure 5.2.1 Illustration of a partitioned matrix, its graph, and its quotient graph

$G(\bigcup_{i=j}^{\ell} L_i)$, each L_j is partitioned according to the sets specified by

(5.3.1) $\{Y \mid Y = L_j \cap C,\ C \text{ a connected component of } G_j\}$

In the example in Figure 5.3.1, the nodes of G_5 are $\{10,15,16,18,19,20\}$ and the corresponding Y's prescribed by (5.3.1) for $j = 5$ are $\{10,15\}$ and $\{18,19\}$.

We now describe an algorithm for finding such a partitioning, given a root r for a rooted level structure. Note that it uses a stack, and thereby avoids explicitly finding the connected components of the G_j. In the algorithm, Span(Y) is the set $\{x \in X \mid \exists \text{ a path from } y \text{ to } x, \text{ for some } y \in Y\}$, where $Y \subseteq X$. If $Y = \{y\}$, Span(y) is simply the connected component containing y.

<u>Step 0</u> (Initialization): Empty the stack. Generate the level structure $\mathcal{L}(r) = \{L_0, L_1, L_2, \ldots, L_{\ell(r)}\}$ rooted at r, and choose any node $y \in L_{\ell(r)}$. Set $\ell = \ell(r)$ and $S \leftarrow \{y\}$.

<u>Step 1</u> (Pop stack): If the stack is empty, set $T = \phi$. Otherwise let T be the node set on the top of the stack. If $T \cap L_\ell \neq \phi$, pop T from the stack and set $S = S \cup T$.

<u>Step 2</u> (Form possible partition member): Determine the set $Y \leftarrow \text{Span}(S)$

in the subgraph $G(L_\ell)$. If some node in $Adj(Y) \cap L_{\ell+1}$ has not yet been placed in a partition member, go to step 5.

Step 3 (New partition member): Put Y in P.

Step 4 (Next level): Determine the set $S \leftarrow Adj(Y) \cap L_{\ell-1}$, and set $\ell \leftarrow \ell-1$. If $\ell \geq 0$, go to step 1, otherwise stop.

Step 5 (Partially formed partition member): Push S onto the stack. Pick $y_{\ell+1} \in Adj(Y) \cap L_{\ell+1}$ and trace a path $y_{\ell+1}, y_{\ell+2}, \ldots, y_{\ell+t}$, where $y_{\ell+i} \in L_{\ell+i}$ and $Adj(y_{\ell+t}) \cap L_{\ell+t+1} \neq \phi$. Let $S \leftarrow \{y_{\ell+t}\}$ and $\ell \leftarrow \ell+t$, and then go to step 1.

The example in Figure 5.3.1, taken from [19], illustrates the results of the algorithm. The level structure rooted at node 1 is refined to obtain a quotient tree having 10 nodes. In the example, $Y_1 = \{20\}$, $Y_2 = \{18,19\}$, $Y_3 = \{16\}$, $Y_4 = \{10,15\}$, $Y_5 = \{9,14,17\}$, and $Y_6 = \{5,11\}$, with $L_4 = Y_5 \cup Y_6$, $L_5 = Y_2 \cup Y_4$, and $L_6 = Y_1 \cup Y_3$.

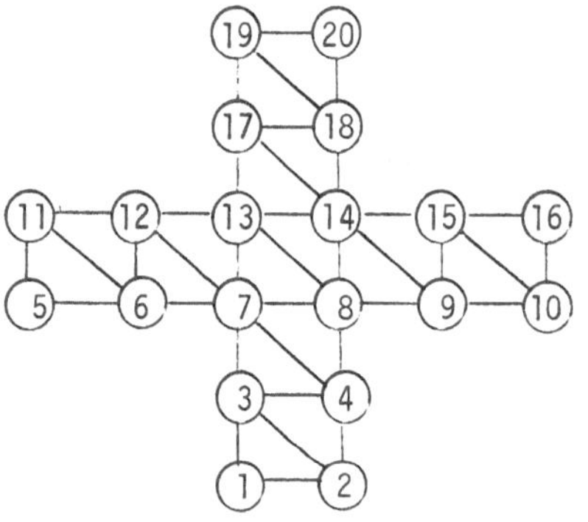

A '+' shaped graph

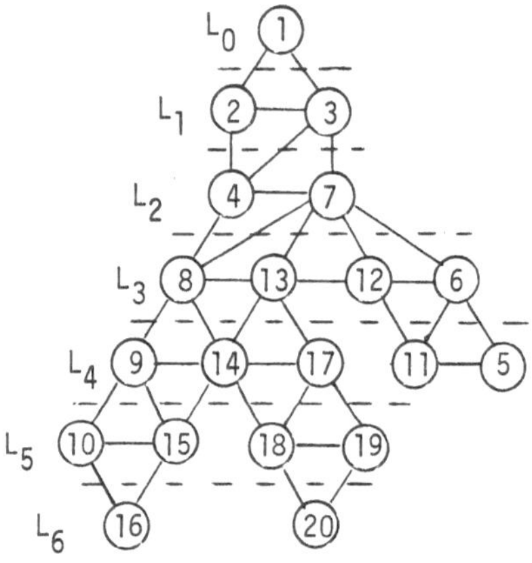

Rooted level structure $\mathcal{L}(1)$

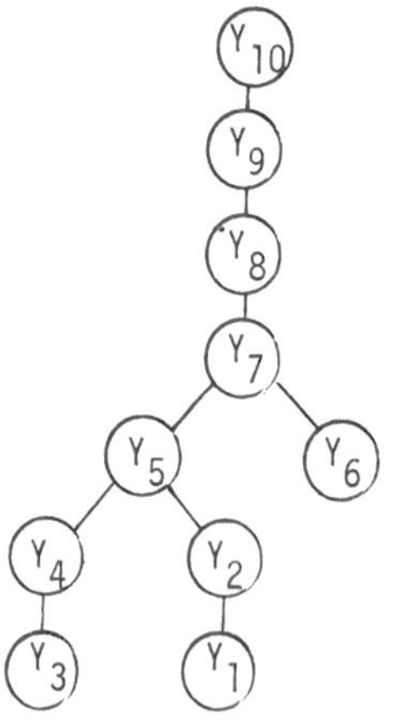

Figure 5.3.1 A graph, rooted level structure, and refined quotient tree

It can be shown that the tree partitioning produced by the algorithm is maximal [19]. Note that the order in which the partition members are found provides a monotone ordering of the quotient tree.

So far we have not specified how to choose the starting node r, and we have not described how the node numbering of each partition member is to be accomplished. Doing this will complete our description of the refined quotient tree (RQT) algorithm.

The node r is obtained by using our modification of the algorithm due to Gibbs, et al., described in section 2.4. Since we want a partitioning with as many members as possible, this seems to be a sensible choice since it will tend to provide a level structure having relatively many levels.

We intend to use the implicit storage scheme described in section 4.3, where the diagonal blocks are stored using Jennings' scheme. Thus, it is obvious that we want to order each partition member so that the diagonal block has a small profile. The numbering scheme is as follows. Let $Y \in P$, where $Y \in L_j$ and P is the refined partitioning of $\mathcal{L}(r) = \{L_0, L_1, \ldots, L_{\ell(r)}\}$. Let S be the set $\{y \in Y | \mathrm{Adj}(y) \cap L_{j+1} \neq \phi\}$. Nodes in the subgraph $G(Y \backslash S)$ are first numbered using the RCM algorithm (described in section 3.3). The nodes in S are then numbered arbitrarily. Note that $G(Y \backslash S)$ may be disconnected.

The set S is the same set S of step 2 of the partition-finding algorithm described above. Thus, it makes sense to carry out the ordering at the same time as the partitioning is found. This can be achieved by inserting the following between steps 3 and 4.

Step 3' (Internal numbering of partition member): Number $G(Y \backslash S)$ using the RCM scheme, and then number S in any order.

The example in Figure 5.3.2, taken from [19], illustrates the effect of this internal numbering scheme. In this particular example no refinement can be done, so the members of P and $\mathcal{L}$ are the same. Note that the nodes associated with each triangle in the mesh are all connected in the corresponding graph.

It is interesting to note the heavy use of the algorithm for finding pseudo-peripheral nodes; it is used to find the starting node for $\mathcal{L}$, as well as the starting node for each application of the RCM algorithm.

5.4 Numerical Experiments

In this section we provide some numerical experiments comparing the quotient tree approach of this section to the envelope scheme of section 3. We denote our implementation of Cholesky's method using Jennings' storage scheme by ENVSLV, and our implementation of Cholesky's method using the implicit storage scheme of section 4.3 by TRESLV. Thus, we are interested in comparing the RCM-ENVSLV combination of algorithms to the RQT-TRESLV combination.

Recall from observation 1 of section 4.2 that we may carry out the factorization in two distinctly different ways, which we referred to as the symmetric and asymmetric forms. We denote the corresponding versions of our TRESLV program by

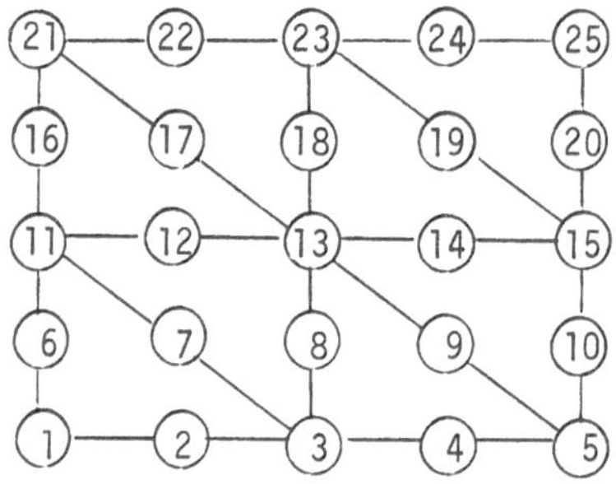

A finite element mesh M

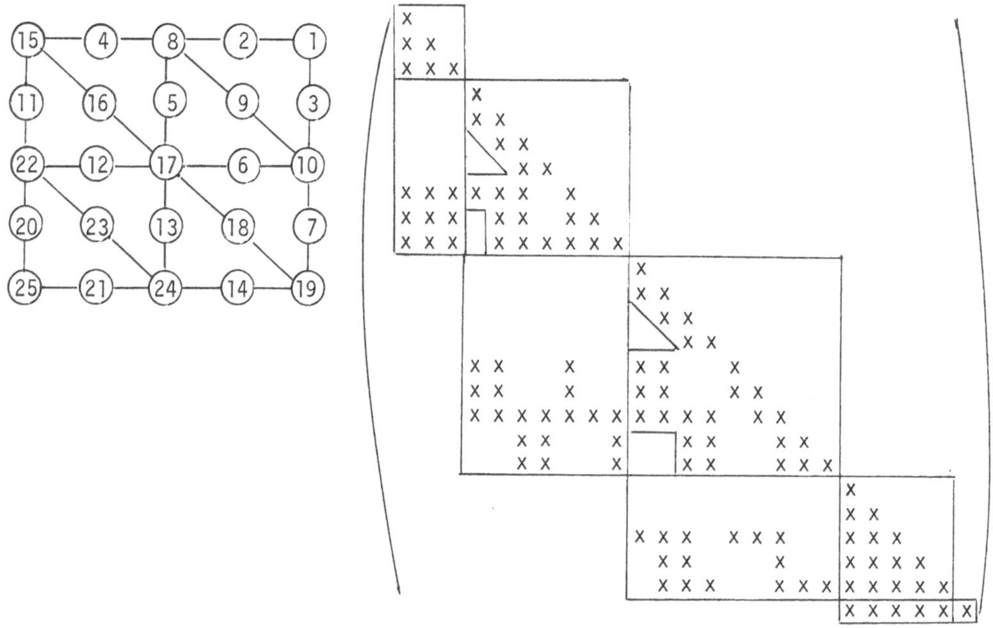

Reordered mesh and its corresponding matrix structure

Figure 5.3.2 An example illustrating the effect of the internal numbering scheme

TRESLV (F_1) and TRESLV (F_2).

Tables 5.4.1 and 5.4.2 summarize the results of our numerical experiments for the twelve test problems considered in section 3. The results demonstrate that for our examples, the F_1 version is significantly more efficient in terms of execution time, and for the low order elements (t=1), required only slightly more storage than the F_2 version. However, for t=2, the F_2 version required substantially less storage. As we mentioned in section 1.3, the computer charging algorithm must be known in order to determine which scheme is most desirable in monetary terms.

TABLE 5.4.1 Factorization and solution times for the test examples

Test Example	Factorization						Solution			
	Operations			Time			Operations		Time	
	RCM ENVSLV	RQT TRESLV(F_1)	RQT TRESLV(F_2)	RCM ENVSLV	RQT TRESLV(F_1)	RQT TRESLV(F_2)	RCM ENVSLV	RQT TRESLV	RCM ENVSLV	RQT TRESLV
a.1	344,608	351,648	564,504	3.39	4.35	6.46	51,106	54,338	.42	.66
a.2	420,827	629,648	788,002	4.27	6.57	8.45	56,894	61,634	.51	.68
b.1	487,992	485,828	791,404	4.52	5.54	8.51	60,056	62,946	.49	.67
b.2	422,357	676,459	845,540	4.30	7.28	9.22	55,952	62,120	.51	.70
c.1	332,412	120,577	189,061	3.09	2.03	2.96	51,720	34,132	.41	.53
c.2	241,666	138,688	181,588	2.23	1.84	2.45	39,724	30,926	.32	.40
d.1	195,063	105,874	164,646	2.15	2.08	2.88	43,364	35,326	.38	.61
d.2	85,875	69,679	93,299	1.00	1.12	1.45	21,992	21,898	.20	.33
e.1	301,788	309,769	493,066	3.07	3.93	5.85	45,506	48,266	.39	.55
e.2	355,862	549,758	692,029	3.51	5.68	7.11	49,578	54,778	.43	.57
f.1	300,226	300,496	485,600	3.24	4.18	6.22	56,436	60,630	.51	.79
f.2	232,144	361,136	457,388	3.67	4.11	5.12	44,400	51,314	.38	.59

TABLE 5.4.2 Ordering times and storage requirements for the examples

Test Example	Order Time		Storage							
			Primary		Overhead			Total		
	RCM	RQT	RCM ENVSLV	RQT TRESLV	RCM ENVSLV	RQT TRESLV(F_1)	RQT TRESLV(F_2)	RCM ENVSLV	RQT TRESLV(F_1)	RQT TRESLV(F_2)
a.1	.30	.35	25,553	14,641	1,092	5,544	5,511	26,645	20,185	20,152
a.2	.44	.73	28,447	16,721	1,092	9,160	5,959	29,539	25,881	22,680
b.1	.28	.33	30,028	16,717	1,012	5,184	5,071	31,040	21,901	21,788
b.2	.41	.68	27,976	16,747	1,012	8,793	5,512	28,988	25,540	22,259
c.1	.38	.45	25,860	9,645	1,183	5,633	6,001	27,043	15,278	15,646
c.2	.43	.70	19,862	8,816	948	6,518	5,111	20,810	15,334	13,927
d.1	.40	.49	21,682	10,124	1,380	6,296	7,023	23,062	16,420	17,147
d.2	.33	.58	10,996	6,381	808	4,872	4,385	11,804	11,253	10,726
e.1	.25	.30	22,753	12,967	939	4,474	4,708	23,692	17,441	17,675
e.2	.39	.66	24,789	14,811	939	7,236	5,092	25,728	22,047	19,903
f.1	.35	.43	28,218	16,526	1,443	6,220	7,240	29,661	22,746	23,766
f.2	.40	.72	22,200	14,173	1,155	6,408	6,228	23,355	20,581	20,401

In general, the use of the quotient tree schemes tended to increase computation time and decrease storage requirements, when compared to the standard envelope scheme. As expected, the quotient tree schemes appear to be most attractive for those problems having appendages; for these, the quotient tree actually has some branching rather than being a simple chain. We note again that a significant part of the total storage requirement for the quotient tree methods is overhead for pointers, so that for computers with a relatively large word size where pointers could be packed several to a word, the total storage for RQT-TRESLV would be reduced relatively more than would the total storage for RCM-ENVSLV.

Finally, it is interesting to note that even for problems as large as 1000 equations, the time spent finding the ordering remains a significant fraction of the overall time used to solve the problem.

Just as we did in section 3 for the RCM-ENVSLV combination of algorithms, we solved several of the mesh problems with increasing mesh subdivision factors in order to gain some insight into the asymptotic behavior of the programs. The results are summarized for the graded-L problem in Tables 5.4.3 and 5.4.4.

As one expects, the ordering time appears to grow linearly with N, and the total storage requirement appears to grow no faster than $O(N^{3/2})$. The TRESLV overhead storage is much more substantial than for the envelope scheme, but again appears to grow only linearly with N. Note the substantial reduction in total storage requirements compared to the envelope storage scheme results in Table 3.4.2, for the same set of problems. The factorization and solution times (and corresponding operation counts) appear to grow as N^2 and $N^{3/2}$ respectively. Note that the F_1-TRESLV factorization program executes slightly fewer arithmetic operations than the ENVSLV program (Table 3.4.2), but because of differences in program complexity, the execution time of the TRESLV program is somewhat larger.

It is interesting to note that if we assume that the computer charging algorithm is the product of execution time and storage requirements, for larger problems the RQT-TRESLV(F_1) program is considerably more efficient than RCM-ENVSLV for this problem, as shown in Table 5.4.5. This was generally the case for all our test problems, but the percentage gain varied over the different domains. The saving was quite spectacular for the + shaped domain, for example. (As an indication, see lines c in Table 5.4.2.) Note that because the TRESLV program involves somewhat more sophisticated data structures than the ENVSLV program, there is a cross-over point below which the TRESLV program is less efficient than the ENVSLV program.

6. Dissection (Substructuring) of Finite Element Problems

6.1 Introduction

In many engineering applications, particularly in the design large and complicated devices, it is convenient to allocate the design of certain components ("substructures") to individual design groups. The modelling of each substructure is carried out more or less independently, and the dependencies between the components

TABLE 5.4.3 Performance statistics on the ordering program and storage scheme for the graded-L domain with s=4(1)12 and t=1

s	Order			Storage (F_1 version of TRESLV)						
	N	Time	$\frac{\text{Time}}{N}$	Primary	$\frac{\text{Primary}}{N^{3/2}}$	Overhead	$\frac{\text{Overhead}}{N}$	$\frac{\text{Overhead}}{\text{Primary}}$	Total	$\frac{\text{Total}}{N^{3/2}}$
4	265	.10	3.77(-4)	2,543	.589	1,381	5.21	.543	3,923	.909
5	406	.14	3.45(-4)	4,606	.563	2,103	5.18	.457	6,709	.820
6	577	.20	3.47(-4)	7,552	.545	2,978	5.16	.394	10,530	.760
7	779	.26	3.34(-4)	11,537	.532	4,005	5.15	.347	15,542	.716
8	1009	.34	3.57(-4)	16,717	.522	5,184	5.14	.310	21,910	.684
9	1270	.41	3.23(-4)	23,248	.514	6,515	5.13	.280	29,763	.658
10	1561	.50	3.20(-4)	31,286	.507	7,998	5.12	.256	39,284	.637
11	1882	.60	3.19(-4)	40,987	.502	9,633	5.12	.235	50,620	.620
12	2233	.71	3.17(-4)	52,507	.498	11,420	5.11	.217	63,927	.606

TABLE 5.4.4 Performance statistics on the linear equations solver for the graded-L domain with s=4(1)12 and t=1

		Factorization (F_1 version of TRESLV)					Solution				
s	N	Time	$\frac{\text{Time}}{N^2}$	Operations	$\frac{\text{Operations}}{N^2}$	$\frac{\text{Time}}{\text{Operations}}$	Time	$\frac{\text{Time}}{N^{3/2}}$	Operations	$\frac{\text{Operations}}{N^{3/2}}$	$\frac{\text{Time}}{\text{Operations}}$
4	265	.61	8.60(-6)	37,050	.528	16.46(-6)	.13	3.01(-5)	9,170	2.13	14.18(-6)
5	406	1.20	7.28(-6)	83,735	.508	14.33(-6)	.22	2.69(-5)	16,872	2.06	13.06(-6)
6	577	2.16	6.49(-6)	164,637	.495	13.12(-6)	.34	2.60(-5)	27,986	2.02	12.86(-6)
7	778	3.64	6.01(-6)	293,370	.485	12.41(-6)	.49	2.35(-5)	43,136	1.99	11.82(-6)
8	1009	5.85	5.74(-6)	485,820	.477	12.04(-6)	.68	2.15(-5)	62,946	1.96	10.96(-6)
9	1270	8.29	5.14(-6)	760,185	.471	10.91(-6)	.89	1.97(-5)	88,040	1.95	10.11(-6)
10	1561	12.57	5.16(-6)	1,136,895	.467	11.06(-6)	1.17	1.90(-5)	119,042	1.93	9.83(-6)
11	1882	16.82	4.75(-6)	1,638,692	.463	10.26(-6)	1.50	1.84(-5)	156,576	1.92	9.58(-6)
12	2233	22.91	4.59(-6)	2,290,590	.459	10.00(-6)	1.89	1.79(-5)	201,266	1.91	9.39(-6)

TABLE 5.4.5 Comparison of envelope and F_1-quotient tree schemes, assuming cost is proportional to storage x execution time, for the graded-L domain with s=4(1)12

s	N	Time				Total Storage		Cost Ratio: quotient tree/envelope	
		Total		Fact. + Solution					
		RCM ENVSLV	RQT TRESLV(F_1)	RCM ENVSLV	RQT TRESLV(F_1)	RCM ENVSLV	RQT TRESLV(F_1)	TOTAL	Factorization + Solution
4	265	.58	.84	.50	.74	4,474	3,923	1.27	1.30
5	406	1.14	1.56	1.03	1.42	8,264	6,709	1.11	1.12
6	577	2.02	2.70	1.86	2.50	13,747	10,530	1.02	1.03
7	778	3.36	4.39	3.15	4.13	21,235	15,542	.96	.96
8	1009	5.16	6.87	4.88	6.53	31,040	21,910	.94	.94
9	1270	7.75	9.59	7.41	9.18	43,474	29,763	.85	.85
10	1561	11.21	14.24	10.79	13.74	58,849	39,284	.85	.85
11	1882	15.73	18.92	15.22	18.32	77,477	50,620	.78	.78
12	2233	21.50	25.51	20.90	24.80	99,670	63,927	.76	.76

resolved after the individual modelling is done. Of course, the interdependencies among the components may require redesign of some of the components, so the above general procedure may be iterated several times.

In order to make these remarks specific, suppose we have a finite element mesh, and we choose a set of mesh vertices S and their incident element edges which, if removed from the mesh, disconnect it into two substructures. (Obviously the same statement can be phrased in terms of finite element graphs, separators, and connected components.) If we number the variables associated with each substructure consecutively, followed finally by the variables associated with S, we induce the following partitioning of the coefficient matrix A.

(6.1.1) $$A = \begin{pmatrix} A_{11} & 0 & A_{13} \\ 0 & A_{22} & A_{23} \\ A_{13}^T & A_{23}^T & A_{33} \end{pmatrix} .$$

The Cholesky factor L of A, correspondingly partitioned, is given by

(6.1.2) $$L = \begin{pmatrix} L_{11} & & \\ 0 & L_{22} & \\ W_{13}^T & W_{23}^T & L_{33} \end{pmatrix} ,$$

where $A_{11} = L_{11}L_{11}^T$, $A_{22} = L_{22}L_{22}^T$, $W_{13} = L_{11}^{-1}A_{13}$, $W_{23} = L_{22}^{-1}A_{23}$, and

(6.1.3) $$L_{33}L_{33}^T = \tilde{A}_{33} = A_{33} - A_{13}^T A_{11}^{-1} A_{13} - A_{23}^T A_{22}^{-1} A_{23} .$$

Here the matrices A_{11} and A_{22} correspond to each component or substructure, and the matrices A_{13} and A_{23} represent the "glue" which relates the substructures through the nodes of S.

In addition to the managerial advantages of substructuring mentioned above, it should be clear to the reader that the partitioned matrix techniques described in section 4.2 can also be applied to the matrix (6.1.1). In addition, since the factors of A_{11} and A_{22} are independent, they can be computed in either order, or in parallel if two processors are available. Finally, in some design applications, several substructures may be identical; for example, the finite element model of sections of a long ship may all be the same, and each may be regarded as a single superelement [2], which is constructed once and used repeatedly in the design of several ships. (In our example above, A_{11} and A_{22} could be identical.)

These dissection or substructuring ideas can be recursively applied, so that each substructure is itself decomposed into smaller substructures, yielding a nested dissection [13]. The orderings thus induced usually lead to very low operation counts and fill. For example, such a dissection ordering of a q×q regular grid can be shown to yield an operation count of $O(q^3)$, compared to the usual $O(q^4)$ for the "natural" row by row ordering. The number of nonzeros in the corresponding factor L is only $O(q^2\log_2 q)$ rather than $O(q^3)$.

However, until fairly recently, the technique of nested dissection (or "multi-level substructuring" or "multi-level superelement analysis") was not widely recognized as a method for inducing low operation count or low-fill orderings. Even when it was recognized as such, practical experience suggested that increased program overhead and storage management requirements outweighed the advantage of low operation or fill counts over more conventional orderings. Thus, these schemes have been generally regarded as techniques to facilitate project management and utilize auxiliary storage, rather than techniques to produce efficient orderings in the sense of arithmetic or fill.

This latter view, however, is precisely the one we intend to adopt in this section. Some recent implementation studies by the author [15] show that with some care in data structure and program design, these nested dissection orderings can be utilized so that program and storage overhead are relatively small. These studies were done for the special q×q grid, and although they established that dissection orderings could be practically useful, the problem of automatically generating such orderings for irregular meshes was not addressed. One of our main objectives here is to describe an algorithm for automatically finding nested dissection orderings.

6.2 Formal Description of Nested Dissection Orderings

In this section we formally define what we mean by a nested dissection ordering, in preparation for a description of an algorithm for producing such orderings, which is the subject of section 6.3. The definition is phrased in terms of a graph, rather than the underlying finite element mesh or matrix.

A nested dissection ordering of a graph $G = (X,E)$ is defined by the following process. Let $R_0 = X$ and for $\ell = 0,1,\ldots$ until $R_\ell = \phi$ do the following:

(i) Determine the connected components of the section graph $G(R_\ell)$ and label the corresponding vertex sets $R_\ell^1, R_\ell^2, \ldots, R_\ell^{r_\ell}$.

(ii) For $j = 1,2,\ldots,r_\ell$ choose $S_\ell^j \subseteq R_\ell^j$ such that S_ℓ^j is either a separator of $G(R_\ell^j)$ or is equal to R_ℓ^j. Define $S_\ell = \bigcup_{j=1}^{r_\ell} S_\ell^j$.

(iii) Define $R_{\ell+1} = R_\ell \setminus S_\ell$.

The partitioning $P = \{S_m^k\}$ of X thus defined is called a nested dissection partitioning (ND-partitioning). A minimal ND-partitioning (MND-partitioning) is one where every separator is minimal. It is helpful to associate with P a rooted dissection tree (S_0,T), where the nodes of T are the members of P, and S_m^k is the father of S^{ℓ}_{m+1} if and only if $R^k_m \supseteq R^{\ell}_{m+1}$. Note that we are using the fact that there is a 1-1 correspondence between the sets $\{R^k_m\}$ and $\{S^k_m\}$. An example illustrating these ideas is contained in Figure 6.2.1. Note that since P is not in general a tree partitioning, the quotient graph G/P is not in general a tree.

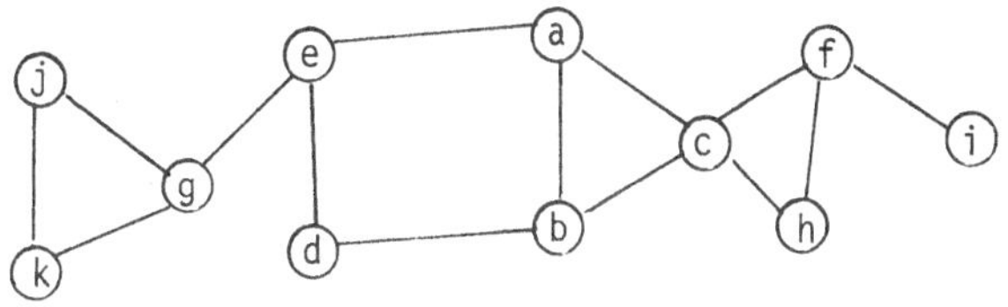

graph G

$R_0 = \{a,b,c,\ldots,k\}$ $\qquad S_0 = \{a,b\}$

$R^1_1 = \{c,f,h,i\}$, $R^2_1 = \{e,d,g,j,k\}$, $S^1_1 = \{f\}$, $S^2_1 = \{e\}$

$R^1_2 = \{d\}$, $R^2_2 = \{g,j,k\}$, $R^3_2 = \{c,h\}$, $R^4_2 = \{i\}$,

$S^j_2 = R^j_2$, $j = 1,2,3,4$. $P = \{S_0, S^1_1, S^2_1, S^1_2, S^2_2, S^3_2, S^4_2\}$

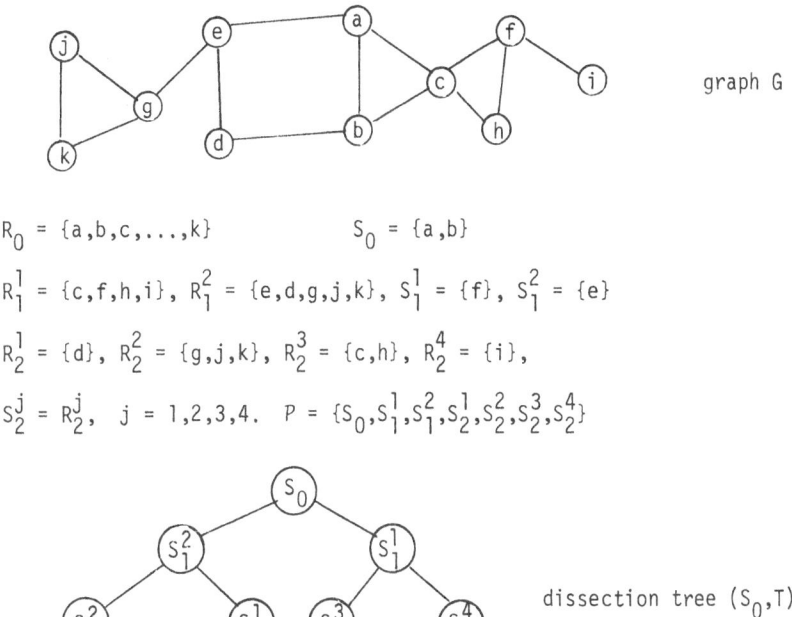

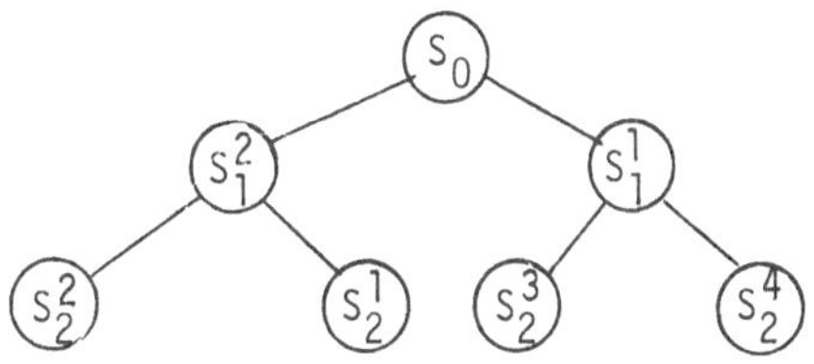

dissection tree (S_0,T)

Figure 6.2.1 A graph G with ND-partitioning and dissection tree (S_0,T)

An ordering α of X is <u>consistent</u> with the ND-partitioning P if and only if $x \in S^j_\ell$ and $y \in R^j_\ell \setminus S^j_\ell \Rightarrow \alpha^{-1}(x) > \alpha^{-1}(y)$. Note that a consistent ordering need not be compatible with the partitioning P; that is, members of P need not be numbered consecutively. Figure 6.2.2 illustrates this point.

Obviously, a monotone ordering of the rooted dissection tree (S_0,T) <u>induces</u> a class of orderings which are both consistent and compatible with P. An ordering α is a <u>nested dissection ordering</u> if it is consistent with some ND-partitioning.

It can be shown (see [20]) that given an ND-partitioning P, it is sufficient to consider only those orderings induced by monotone orderings of the dissection tree. Moreover, all such monotone orderings are equivalent in the sense that least-fill orderings induced by monotone orderings of the dissection tree all produce the same fill and operation count.

Let $\eta(G^\alpha)$ be the number of nonzeros in the triangular factor L of A, where the graph G of A is ordered according to α. The following two theorems have been proved in [20].

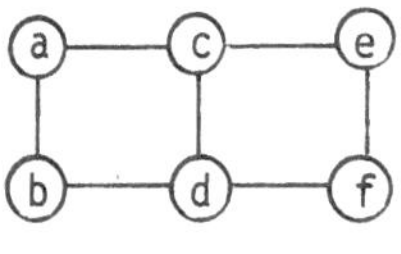

$S_0 = \{c,d\}$

$S_1^1 = \{a,b\}, \quad S_1^2 = \{e,f\}$

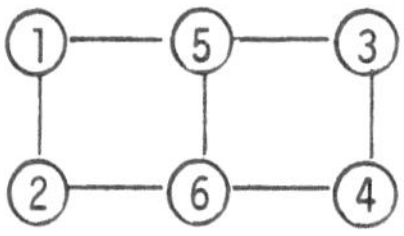

Ordering consistent and compatible with P

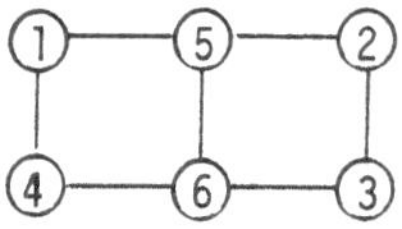

Ordering consistent with P but not compatible with P

Figure 6.2.2 An example illustrating consistency and compatibility of an ordering with respect to an ND-partitioning

Theorem 6.2.1

Let $P = \{S_m^k\}$ be an ND-partitioning of $G = (X,E)$ and let α be an ordering consistent with P. Then

$$\eta(G^\alpha) \le \sum_{\ell=0}^{h} \sum_{j=1}^{r_\ell} |S_\ell^j| \{ |Adj(R_\ell^j)| + (|S_\ell^j|-1)/2 \}, \; h = \max\{\ell \mid R_\ell \ne \phi\}.$$

The assumptions in the following corollary are motivated by the class of two dimensional finite element problems we use as test problems.

Corollary 6.2.2

Let P and α be as in Theorem 6.2.1, and suppose further that P satisfies the following properties, where $N = |X|$.

1) $h \le C_1 \log_2 N$

2) $|S_\ell^j| \le C_2 (N/2^\ell)^{1/2}$

3) $r_\ell \le C_3 \cdot 2^\ell$

4) $|Adj(R_\ell^j)| \le C_4 |S_\ell^j|$.

Then

$$\eta(G^\alpha) \le C_5 N \log_2 N$$

Here the C_i are constants, independent of N.

The above theorem and corollary suggest a recursive application of the following heuristic: find a "small" separator which disconnects the graph into two or more components of approximately equal size. The algorithm of the next section is essentially an implementation of this heuristic.

6.3 A Heuristic Algorithm for Automatic Nested Dissection

It should be clear from section 6.2 that a major aspect of the ordering algorithm will involve finding the separators S_ℓ^j of the subgraphs $G(R_\ell^j)$. The algorithm we propose makes heavy use of the algorithm for finding pseudo-peripheral nodes described in section 2.4, along with the concept of rooted level structures, defined in section 2.2 and used extensively in the algorithm for finding tree partitionings described in section 5.3.

Loosely speaking, the algorithm finds a separator of a graph (or subgraph) by first generating a level structure of the graph rooted at a pseudo-peripheral node, and then choosing a subset of the "middle" level of the level structure which is a minimal separator. We root the level structure at a pseudo-peripheral node because this will tend to produce a structure having relatively many levels, and the levels will tend to have relatively few members.

A complete description of the algorithm is as follows:

Automatic Nested Dissection Algorithm

1. Set $P = \{\phi\}$, $C = \phi$, and $N = |X|$.
2. If $N = 0$, stop. Otherwise find a pseudo-peripheral node y in some connected component of $G(X \backslash C)$, say $G(R_m^k)$.
3. Generate the rooted level structure $\mathcal{L}(y) = \{L_0, L_1, L_2, \ldots, L_{\ell(y)}\}$ spanning the subgraph $G(R_m^k)$. If $\ell(y) \leq 1$, set $S_m^k = R_m^k$ and go to 5. Otherwise set $j = \lfloor(\ell(y)+1)/2\rfloor$.
4. Choose $S_m^k \subseteq L_j \ni S_m^k$ is a minimal separator.
5. Set $C \leftarrow C \cup S_m^k$ and $P = P \cup \{S_m^k\}$. Label the nodes of S_m^k from $N-|S_m^k|+1$ to N, using the RCM algorithm applied to the subgraph $G(S_m^k)$.
6. Set $N \leftarrow N-|S_m^k|$ and go to step 2.

There are two reasons for the use of the RCM algorithm in step 5 to number the partition members, both motivated by the fact that we intend to use the block-within-block-column storage scheme described in section 4.3. First, when $S_m^k = R_m^k$, so that S_m^k is a pendant node of the dissection tree, the envelope of the diagonal block of the correspondingly ordered matrix will be preserved during the factorization. Since we use Jennings' storage scheme for the diagonal blocks, ordering these pendant node sets using the RCM algorithm will tend to reduce primary storage requirements.

When $S_m^k \neq R_m^k$, but is instead a minimal separator, it can be shown [20] that the corresponding diagonal block will completely fill in, so the application of the RCM algorithm to these node sets cannot be justified on the basis of primary storage reduction. However, its use does in general reduce overhead storage substantially when compared to an arbitrary ordering of the S_m^k. This can be explained as follows. In general, the non-pendant nodes of the dissection tree correspond to node sets lying on edge sequences of the finite element mesh. If these nodes are numbered consecutively, beginning at one end of the edge sequence (as the RCM algorithm described in section 3.3 will do), the non-null rows within each block column of L will be bunched together as blocks. Since the overhead storage increases with the number of off-diagonal blocks, this arrangement of the non-null rows will reduce the overhead storage. Figure 6.3.1 contains an example illustrating this point. The non-null rows of block-columns 1 through 4 are grouped together for the ordering α_1, but not for α_2.

6.4 Numerical Experiments

In this section we present some numerical experiments demonstrating the performance of the automatic nested dissection algorithm of the previous section. In order to gain some insight into the expected asymptotic behavior of the ordering algorithm and the execution times of the solver, we again solved some of our test problems with various subdivision factors s. Just as we found in section 3.4, the results were similar for each problem, so we again present results only for the graded-L.

Several aspects of the information in Tables 6.4.1 and 6.4.2 are of interest:

1) The execution time of the ordering algorithm appears to be proportional to N log N. This is typical of such "divide and conquer" algorithms [1].

2) The storage overhead appears to grow linearly with N. This contrasts with many sparse matrix storage schemes, where storage overhead is proportional to the number of nonzeros in L. Here it seems clear that overhead storage ÷ primary storage $\rightarrow 0$ as $N \rightarrow \infty$.

3) Although we believe that the operation counts for the factorization grow as $N^{3/2}$, it is difficult to be sure from the tables that this is indeed the case. About all one can say is that the operation counts appear to grow as f(N), where $C_1 N^{3/2} \leq f(N) \leq C_2 N^{3/2} \log N$. Similarly, although it seems plausible from what is known about model problems [13] that the primary storage (hence also the solution operation count), is proportional to N log N, from our tables it seems only safe to conclude that primary storage grows as g(N), where $C_3 N \log N \leq g(N) \leq C_4 N (\log N)^2$. Here the C_i are constants.

4) Even for problems of quite large size, the execution time required for the ordering algorithm is a significant portion of the overall time required to solve the problem. However, as mentioned earlier, there are circumstances where ordering time can be ignored because numerous problems with the same structure must be solved.

Again, assuming a computer charging algorithm which is proportional to the product of execution time and the storage used, we can compare the envelope and dissection schemes for the graded-L problem, as shown in Table 6.4.3. Unlike the previous algorithm combinations considered,the ordering time is a substantial portion of the total time required to solve the problem. Again we have cross-over points

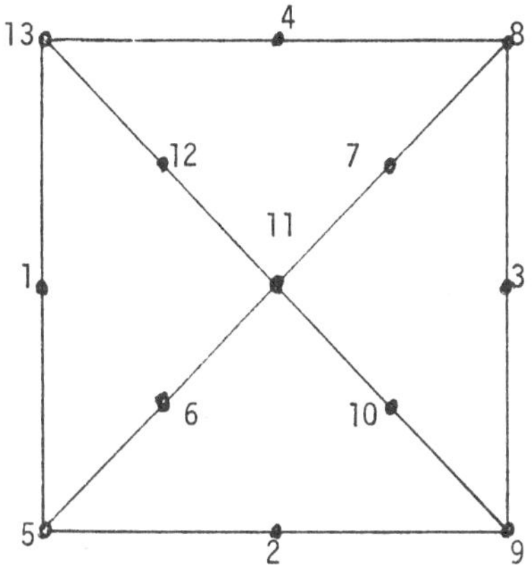

Ordering α_1 for the mesh

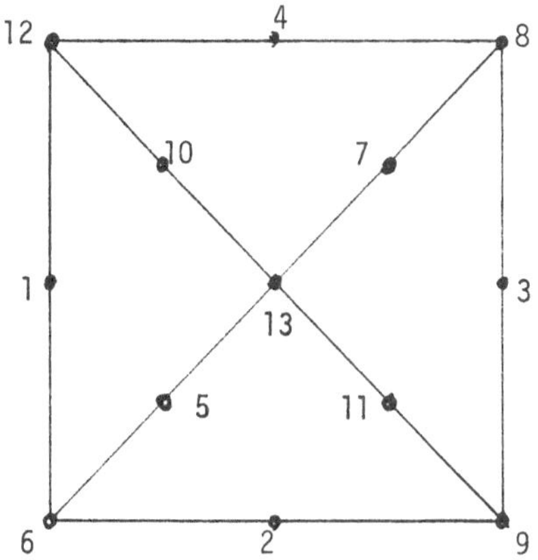

Ordering α_2 for the mesh

$S_0 = \{9,10,11,12,13\}$, $S_1^1 = \{5,6\}$, $S_1^2 = \{7,8\}$

$S_2^1 = \{1\}$, $S_2^2 = \{2\}$, $S_2^3 = \{3\}$, $S_2^4 = \{4\}$

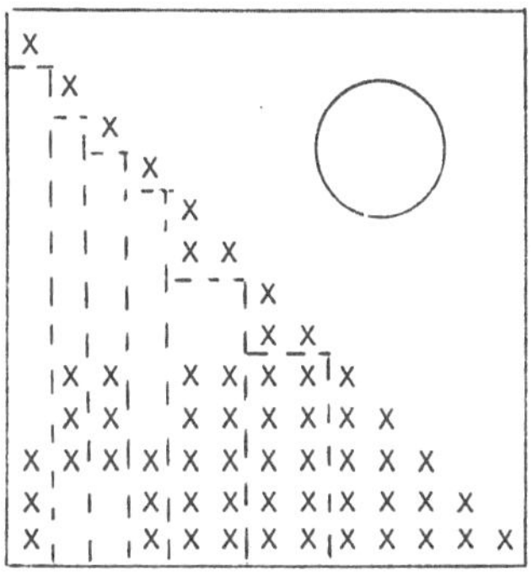

Matrix L corresponding to α_1

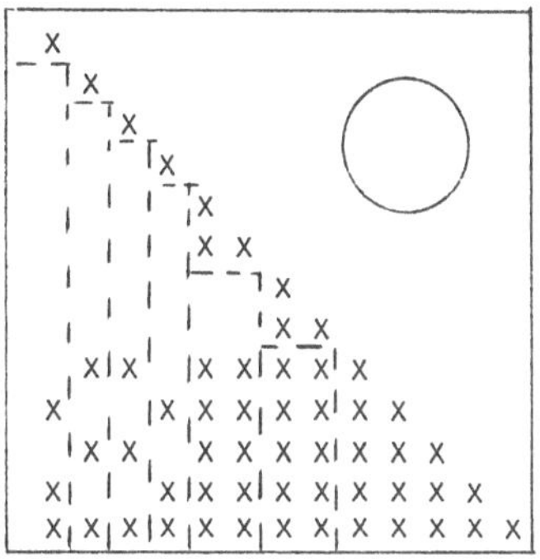

Matrix L corresponding to α_2

Figure 6.3.1 An example illustrating the effect of the ordering of the partitions on the number of blocks within each block column

TABLE 6.4.1 Performance statistics on the ordering program and storage scheme for the graded-L domain with s=4(1)14 and t=1

s	N	Order		Storage					
		Time	Time / N log N	Primary	Primary / N log N	Primary / $N(\log N)^2$	Overhead	Overhead / N	Overhead / Primary
4	265	.71	4.80(-4)	3,671	2.48	.445	1,505	5.68	.410
5	406	1.13	4.63(-4)	6,339	2.60	.433	2,296	5.65	.362
6	577	1.67	4.56(-4)	9,867	2.69	.423	3,317	5.75	.336
7	778	2.39	4.61(-4)	14,321	2.76	.415	4,470	5.75	.333
8	1009	3.06	4.38(-4)	19,846	2.84	.411	5,815	5.76	.293
9	1270	3.94	4.34(-4)	26,313	2.90	.406	7,270	5.73	.276
10	1561	4.96	4.32(-4)	33,904	2.95	.402	8,866	5.68	.262
11	1882	6.10	4.30(-4)	42,586	3.00	.398	10,864	5.77	.255
12	2233	7.39	4.29(-4)	52,524	3.05	.396	12,923	5.79	.246
13	2617	9.07	4.40(-4)	63,350	3.08	.391	15,222	5.81	.240
14	3025	10.51	4.33(-4)	75,553	3.12	.389	17,433	5.76	.231

TABLE 6.4.2 Performance statistics for the linear equations solver for the graced-L domain with s=4(1)14 and t=1

		Factorization						Solution			
s	N	Time	$\frac{\text{Time}}{N^{3/2}}$	$\frac{\text{Time}}{\text{Operations}}$	Operations	$\frac{\text{Operations}}{N^{3/2}}$	$\frac{\text{Operations}}{N^{3/2}\log N}$	Time	$\frac{\text{Time}}{N\log N}$	$\frac{\text{Time}}{\text{Operations}}$	Operations
4	265	.68	1.58(-4)	1.99(-5)	34,096	7.90	1.42	.19	1.29(-4)	2.59(-5)	7,340
5	406	1.21	1.48(-4)	1.75(-5)	68,954	8.43	1.40	.31	1.27(-4)	2.44(-5)	12.676
6	577	2.00	1.44(-4)	1.63(-5)	122,783	8.86	1.39	.45	1.23(-4)	2.28(-5)	19,732
7	778	2.98	1.37(-4)	1.50(-5)	198,984	9.17	1.38	.62	1.20(-4)	2.16(-5)	28,640
8	1009	4.32	1.35(-4)	1.41(-5)	305,729	9.54	1.38	.81	1.16(-4)	2.04(-5)	39,690
9	1270	5.88	1.30(-4)	1.33(-5)	440,803	9.74	1.36	1.04	1.15(-4)	1.98(-5)	52,624
10	1561	7.87	1.28(-4)	1.28(-5)	614,031	9.96	1.35	1.28	1.12(-4)	1.89(-5)	67,806
11	1882	10.31	1.26(-4)	1.25(-5)	821,916	10.14	1.35	1.58	1.11(-4)	1.86(-5)	85,170
12	2233	13.23	1.25(-4)	1.21(-5)	1,091,884	10.35	1.34	1.94	1.13(-4)	1.85(-5)	105,048
13	2617	16.40	1.22(-4)	1.18(-5)	1,394,315	10.42	1.32	2.27	1.10(-4)	1.79(-5)	126,698
14	3025	20.51	1.23(-4)	1.17(-5)	1,757,909	10.57	1.32	2.66	1.10(-4)	1.76(-5)	151,104

TABLE 6.4.3 Comparison of envelope and dissection schemes, assuming cost is proportional to storage x execution time. The problem is the graded-L with t=1 and s=4(1)12

s	N	Time				Total Storage		Cost Ratio	
		Total		Fact. + Solution					
		Envelope	Dissection	Envelope	Dissection	Envelope	Dissection	Total	Fact. + Solution
4	265	.58	1.58	.50	.87	4,474	5,376	3.27	2.09
5	406	1.14	2.65	1.03	1.52	8,264	8,635	2.43	1.54
6	577	2.02	4.12	1.86	2.45	13,747	13,184	1.96	1.26
7	778	3.36	5.99	3.15	3.60	21,235	18,791	1.58	1.01
8	1009	5.16	8.19	4.88	5.13	31,040	25,661	1.31	.87
9	1270	7.75	10.86	7.41	6.92	43,474	33,583	1.08	.72
10	1561	11.21	14.11	10.79	9.15	58,849	42,770	.91	.62
11	1882	15.73	17.99	15.22	11.89	77,477	53,450	.78	.54
12	2233	21.50	22.56	20.90	15.17	99,670	65,447	.69	.48

due to execution overhead and storage overhead.

7. Concluding Remarks

We have described three quite different general strategies for solving finite element systems of equations, and reported numerical experiments for the implementations of each method. Looking at the three algorithms in perspective leads us to the following observations, some of which were already made in section 1.

1. The comparison of two different approaches to solving sparse systems of equations depends very much on individual circumstances such as computer charging algorithms, how many problems having the same structure must be solved, etc.

2. The computer implementation of the algorithm should not be ignored in comparisons of algorithms. Some data structures exact a higher price in execution and storage overhead than others; any comparison which claims to be practical (or relevant) should consider the computer implementation.

3. The use of partitioned matrices in the solution of finite element equations appears to be a very useful tool, particularly in the design of efficient storage schemes. In this connection, step 5 of the nested dissection ordering algorithm illustrates the fact that at least part of the success of an ordering algorithm may depend upon the algorithm "knowing" about the data structure to be used. It is often the case that for a given equation solver, "equivalent" orderings in the sense of operation counts and/or fill incur vastly different execution and storage overheads.

4. Although the three general ordering/solution strategies are quite different, they utilize several common ideas. All three algorithms utilize pseudo-peripheral nodes, and use the algorithm of Gibbs, et al. to find such nodes. Both the quotient tree algorithm and the nested dissection algorithm utilize rooted level structures, pseudo-peripheral nodes, and the RCM algorithm in a subsidiary role. In the same vein, the data structures for partitioned matrices, and the linear equation solver programs which use them,utilize the basic envelope solver ENVSLV in a subsidiary fashion.

This utilization of common ideas in programs illustrates how simple basic techniques can be combined in a hierarchical manner to yield more powerful techniques.

8. References

[1] A.V. Aho, J.E. Hopcroft and J.D. Ullman, The Design and Analysis of Computer Algorithms, Addison-Wesley, Reading, Mass. (1974).

[2] P.O. Araldsen, "The application of the superelement method in the analysis and design of ship structures and machinery components", National Symposium of Computerized Structural Analysis and Design, George Washington University, March 1971.

[3] I. Arany, W.F. Smyth and L. Szoda, "An improved method for reducing the bandwidth of sparse symmetric matrices", in Information Processing 71: Proceedings of IFIP Congress, North-Holland, Amsterdam, 1972.

[4] C. Berge, The Theory of Graphs and its Applications, John Wiley & Sons Inc., New York, 1962.

[5] G. Birkhoff and Alan George, "Elimination by nested dissection", in Complexity of Sequential and Parallel Numerical Algorithms, edited by J.F. Traub, Academic Press, New York and London (1973).

[6] H.L. Crane, Jr., N.E. Gibbs, W.G. Poole, Jr., and P.K. Stockmeyer, "Matrix bandwidth and profile minimization", Report No.75-9, ICASE Report (1975).

[7] E. Cuthill, "Several strategies for reducing the bandwidth of matrices", in Sparse Matrices and Their Applications, D. Rose and R. Willoughby, eds., Plenum Press, New York (1972).

[8] E. Cuthill and J. McKee, "Reducing the bandwidth of sparse symmetric matrices", Proc. 24th Nat. Conf., Assoc. Comput. Mach., ACM Publ. P-69, 1122 Ave. of the Americas, New York, N.Y. (1969).

[9] I.S. Duff, A.M. Erisman and J.K. Reid, "On George's nested dissection algorithm", SIAM J. Numer. Anal., to appear.

[10] G.C. Everstine, "The BANDIT computer program for the reduction of matrix bandwidth for NASTRAN", NSRDC Report 3827 (1972).

[11] Carlos A. Felippa, "Solution of linear equations with skyline-stored symmetric matrix", Computers and Structures, 5 (1975), pp.13-29.

[12] Alan George, "Computer implementation of the finite element method", Stanford Computer Science Dept., Technical Report STAN-CS-71-208, Stanford, California (1971).

[13] Alan George, "Nested dissection of a regular finite element mesh", SIAM J. Numer. Anal., 10 (1973), pp.345-363.

[14] Alan George, "On block elimination for sparse linear systems", SIAM J. Numer. Anal., 11 (1974), pp.585-603.

[15] Alan George, "Numerical experiments using dissection methods to solve n by n grid problems", SIAM J. Numer. Anal., to appear.

[16] Alan George and Joseph W.H. Liu, "A note on fill for sparse matrices", SIAM J. Numer. Anal., 12 (1975), pp.452-455.

[17] Alan George and Joseph W.H. Liu, "An automatic partitioning and solution scheme for solving large sparse positive definite systems of linear algebraic equations", Research Report CS-75-17, Dept. of Computer Science, University of Waterloo, Waterloo, Ontario, Canada (1975).

[18] Alan George, "Sparse matrix aspects of the finite element method", Proc. 2nd International Symposium on Computing Methods in Applied Science and Engineering, Springer-Verlag (1976).

[19] Alan George and Joseph W.H. Liu, "Algorithms for matrix partitioning and the numerical solution of finite element systems", Report CS-76-30, Dept. of Computer Science, University of Waterloo, Waterloo, Ontario, Canada (1976).

[20] Alan George and Joseph W.H. Liu, "An algorithm for automatic nested dissection and its application to general finite element problems", Report CS-76-38, Dept. of Computer Science, University of Waterloo, Waterloo, Ontario, Canada (1976).

[21] N.E. Gibbs, W.G. Poole and P.K. Stockmeyer, "An algorithm for reducing the bandwidth and profile of a sparse matrix", SIAM J. Numer. Anal., 13 (1976), pp.236-250.

[22) F.G. Gustavson, "Some basic techniques for solving sparse systems of equations", Sparse Matrices and their Applications, D.J. Rose and R.A. Willoughby eds., Plenum Press, New York (1972).

[23] A. Jennings, "A compact storage scheme for the solution of simultaneous equations", Comput. J., 9 (1966), pp.281-285.

[24] Joseph W.H. Liu and Andrew H. Sherman, "Comparative analysis of the Cuthill-McKee and the reverse Cuthill-McKee ordering algorithms for sparse matrices", SIAM J. Numer. Anal., 13 (1975), pp.198-213.

[25] Joseph W.H. Liu, "On reducing the profile of sparse symmetric matrices", Rept. CS-76-07, Dept. of Computer Science, University of Waterloo, Waterloo, Ontario, Canada (February 1976).

[26] Christian Meyer, "Solution of linear equations - State-of-the-art", J. of the Struct. Div., ASCE, Proc. Paper 9861 (July 1973), pp.1507-1527.

[27] S.V. Parter, "The use of linear graphs in Gauss elimination", SIAM Rev., 3 (1961), pp.364-369.

[28] W.C. Rheinboldt and C.K. Mestztenyi, "Programs for the solution of large sparse matrix problems based on the arc-graph structure", Technical Report TR-262 (1973), Computer Science Centre, University of Maryland.

[29] Gilbert W. Strang and George J. Fix, An Analysis of the Finite Element Method, Prentice-Hall, Inc., Englewood Cliffs, N.J. (1973).

[30] K.L. Stewart and J. Baty, "Dissection of structures", J. Struct. Div., ASCE, Proc. paper No.4665 (1966), pp.75-88.

[31] James H. Wilkinson, The Algebraic Eigenvalue Problem, Clarendon Press, Oxford (1965).

[32] O.C. Zienkiewicz, The Finite Element Method in Engineering Science, McGraw-Hill, London (1970).

SOLUTION OF LINEAR SYSTEMS OF EQUATIONS: DIRECT METHODS (GENERAL)

John K. Reid

Computer Science and Systems Division
AERE Harwell, Bldg. 8.9
Oxfordshire, OX11 ORA England

CONTENTS

1. Introduction

We consider here the direct solution of sets of linear equations

$$Ax = b \tag{1.1}$$

whose matrix is sparse, without making any assumptions about the nature of the sparsity. This problem has generated a large amount of research in recent years. We will not attempt to survey it comprehensively since Duff (1976a) has written an excellent literature survey including a very good bibliography. We have decided to be selective in our choice of material, which must inevitably mean important omissions. We include only those methods which are of practical importance and which we believe to be at least as good as others currently available. Usually we include descriptions of sufficient detail for this paper to be read on its own, but where the literature contains a really clear description we refer the reader to this in order to save space.

Our attention is given mainly to the case where A is an unsymmetric non-singular nxn matrix, but we also include remarks on how rank deficiency may be treated and discuss the symmetric case and the least squares solution of over-determined linear equations.

A matrix is said to be sparse if it has sufficiently many zero elements for it to be worthwhile to use special techniques that avoid storing or operating with the zeros. Unfortunately, these techniques almost always imply an overhead on other operations so they are worthwhile only when the number of zeros is large. It is almost as difficult to answer the question "when is a matrix sparse?" as "how long is a length of string?", but we will try to give some guidance on this matter in the course of this paper. Sometimes it is obvious that sparse techniques are or are not appropriate. For instance if a 100x100 matrix has ten zeros it is clearly best to store these zeros explicitly and work as if with a full matrix while if a matrix of order 1000 has about 4000 non-zeros it is obviously best to store only the non-zeros and use special coding for handling them.

Matrices of this kind arise in such diverse fields as management science, power systems analysis, surveying, circuit theory, chemical kinetics and structural analysis. That management science gives rise to sparse systems is easy to see, for to describe the operations of a large company might need several thousand variables, and yet individual constraints, such as that on the production of a single factory, involve only a very few variables. In the other fields that I have mentioned the sparsity usually arises because the underlying problem consists of a network. The equation associated with an individual node will involve the variable or variables belonging to that node and the variables belonging to nodes to which it is connected in the network. The matrix is sparse because each node is likely to be connected only to a few neighbouring nodes.

2. Graphs associated with matrices

It is often useful to reverse this process and think of the network, or graph, associated with an arbitrary sparse matrix. This has a node associated with each row and if a_{ij} is non-zero there is a connection from node i to node j in the graph. In the case of a symmetric matrix there would always be a connection from i to j whenever there is one from j to i so we may simplify matters by using undirected connections (lines without arrows). Two simple cases are illustrated in figures 2.1 and 2.2. The directed graph associated with an unsymmetric matrix is usually called a digraph and the word graph is used for the undirected graph associated with a symmetric matrix. A less trivial example arising from the 5-point finite-difference discretization of Laplace's equation is shown in figure 2.3.

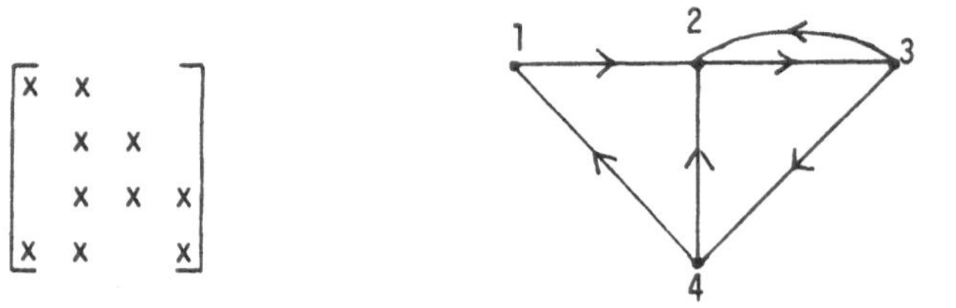

Figure 2.1 An unsymmetrix matrix and its associated digraph.

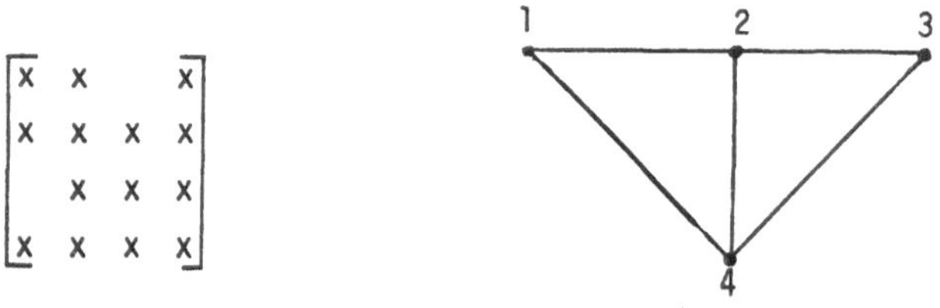

Figure 2.2 A symmetrix matrix and its associated graph.

$$\begin{bmatrix}
-4 & 1 & & & 1 \\
1 & -4 & 1 & & & 1 \\
& 1 & -4 & 1 & & & 1 \\
& & 1 & -4 & & & & 1 \\
1 & & & & -4 & 1 & & & 1 \\
& 1 & & & 1 & -4 & 1 & & & 1 \\
& & 1 & & & 1 & -4 & 1 & & & 1 \\
& & & 1 & & & 1 & -4 & & & & 1 \\
& & & & 1 & & & & -4 & 1 & & & 1 \\
& & & & & 1 & & & 1 & -4 & 1 & & & 1 \\
& & & & & & 1 & & & 1 & -4 & 1 & & & 1 \\
& & & & & & & 1 & & & 1 & -4 & & & & 1 \\
& & & & & & & & 1 & & & & -4 & 1 \\
& & & & & & & & & 1 & & & 1 & -4 & 1 \\
& & & & & & & & & & 1 & & & 1 & -4 & 1 \\
& & & & & & & & & & & 1 & & & 1 & -4
\end{bmatrix}$$

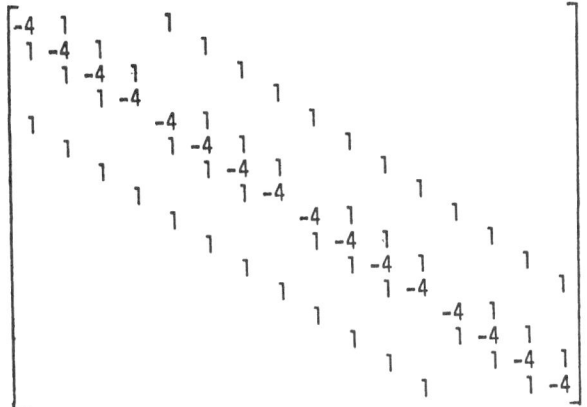

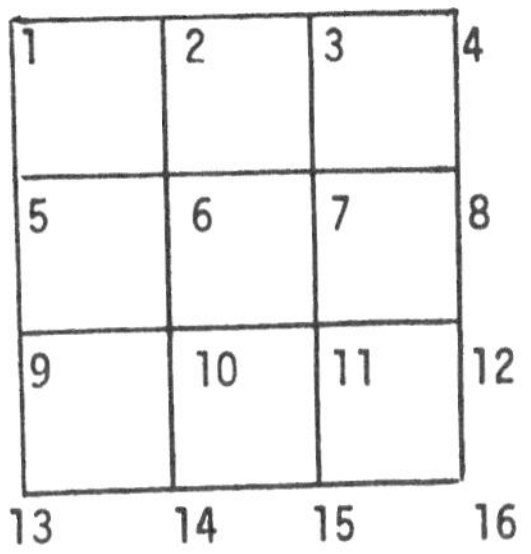

Figure 2.3 **The matrix and associated graph arising from the finite difference approximation of Laplace's equation in a square.**

Direct methods lead to additional non-zeros (fill-ins) being created during solution, and their success depends critically on how many such fill-ins take place. Generally speaking, there are fewer fill-ins in cases where the associated graph can be separated into disjoint parts by the removal of a small number of nodes. Thus matrices arising from power transmission networks are likely to suffer little fill-in whereas those from partial differential equations, particularly in three dimensions, suffer much fill-in.

3. Storing and manipulating sparse matrices

The matrix shown in figure 2.3 has an obvious repeating structure which could be exploited in order to store it in compact form, but in general we need to store each non-zero explicitly along with some indexing information to indicate where it belongs in the matrix. The most convenient way to do this when specifying a matrix is to store just the non-zeros as triples (a_{ij}, i, j) held contiguously in any order in a real array and two integer arrays. Unfortunately, when manipulating it (for example, performing Gaussian elimination) we are likely to require access to the rows and/or the columns, each of which can only be found by scanning the whole stored matrix.

The simplest way to provide such access is by holding two more integers with each non-zero to give the address of the next non-zero in the row and the address of the next non-zero in the column, that is to expand the structure into a linked list. Pointers to the first elements in the rows and columns will also be needed.

Non-zeros may be inserted and deleted easily from such a structure since we need only to adjust the pointers. Storage positions freed by removal of non-zeros may be linked together for reuse later; when a non-zero is added it either uses a space in the free list or, if this is empty, is placed in the next unused space in the arrays.

A full array is often used to hold a sparse vector because such operations as adding a multiple of one component to another can then be applied easily. Alternatively we may hold sets of real-integer pairs (x_i,i) in two arrays either contiguously or as a linked list with an additional integer held with each non-zero to point to the next non-zero. Until recently the author thought that it was usually necessary to hold the non-zeros in order so that two vectors can be scanned in phase when taking their inner product or adding a multiple of one to the other. However this is not necessary if a full vector of workspace w is available. Assuming w is already set to zero we may, for instance, add α times the packed vector y to the packed vector x as follows:

i) Scan y, setting $w(i)=y_i$ for each pair (y_i,i).

ii) Scan x, replacing each (x_i,i) by $(x_i+\alpha*w(i),i)$ and setting w(i) to zero.

iii) Scan y, adding the new non-zero $(\alpha*w(i),i)$ to x and setting $w(i)=0$ for each pair (y_i,i) such that $w(i)\neq 0$.

Returning now to the storage of matrices, we mention that an alternative to the linked list is to hold a file containing the rows and/or columns as packed vectors. If access by rows only is required then we may hold each row as a contiguous set of pairs (a_{ij},j) in a real and an integer array; for each row we need to record its number of non-zeros and the address of its start. This yields a very economical storage scheme, with an overhead of only one integer per non-zero, in contrast to the four that the linked list of the next-to-last paragraph requires. Such a scheme does not provide access to the columns, but Gustavson (1972) has pointed out that adding a file containing the row numbers by columns is sufficient in all likely applications (see section 13, for instance). This file has exactly the same format as that for the packed row vectors except that the real values are omitted. It increases the overhead to two integers per non-zero.

A disadvantage of not using a linked list for storing a matrix is that additional non-zeros cannot be inserted so easily. It is usual to make a fresh copy of the row or column and temporarily to waste the space that the old version occupied. We therefore need some "elbow room" in our arrays and to "compress" the file when there is no longer space for a new vector. Both Zlatev and Thomsen (1976), who have implemented a subroutine for solving sparse sets of equations that uses this storage pattern, and Reid (1976), who has implemented such a scheme for sparse linear programming bases, report that the storage and computing overheads involved are quite moderate.

The storage disadvantage of the linked list was avoided by Curtis and Reid (1971a) by holding only the links and not the row and column numbers. The latter were obtained by following the links to the row or column end where the link was replaced by the row or column number. While this is satisfactory for very sparse cases, it has proved rather costly in cases with a reasonably high (say ten or more) average number of non-zeros per row.

4. Gaussian elimination

Direct methods for solving sets of n linear equations

$$Ax = b \tag{4.1}$$

are variants of the method of Gaussian elimination, which may be summarized by the formulae

$$a_{ij}^{(k+1)} = a_{ij}^{(k)} - a_{ik}^{(k)} [a_{kk}^{(k)}]^{-1} a_{kj}^{(k)} \quad (i,j>k) \tag{4.2}$$

and

$$b_i^{(k+1)} = b_i^{(k)} - a_{ik}^{(k)} [a_{kk}^{(k)}]^{-1} b_k^{(k)} \quad (i>k), \tag{4.3}$$

which express the operations performed at the kth step, $k=1,2,\ldots,n-1$, beginning with $A^{(1)}=A$, $b^{(1)}=b$. This leads eventually to an upper triangular system which can be solved by the back-substitution

$$x_k = [a_{kk}^{(k)}]^{-1} \left(b_k^{(k)} - \sum_{j=k+1}^{n} a_{kj}^{(k)} x_j\right) , \quad k=n,n-1,\ldots,1. \tag{4.4}$$

Each $a_{ij}^{(k+1)}$ may over-write $a_{ij}^{(k)}$ in storage and the "multipliers"

$\ell_{ik} = a_{ik}^{(k)} [a_{kk}^{(k)}]^{-1}$ may overwrite $a_{ik}^{(k)}$. We will have obtained the triangular factorization

$$A = L\,U \tag{4.5}$$

where L is unit lower triangular with off-diagonal elements ℓ_{ik} and U is the upper triangular matrix $\{a_{ij}^{(i)},\ i\le j\}$.

5. Numerical stability of Gaussian elimination

Numerical instability is associated with large numbers being added to small ones in equation (4.2), for this may mean that information present in the small number is lost. For example in 4-decimal arithmetic we find

$$1.234 + 402.1 = 403.3 \tag{5.1}$$

and the last two digits of 1.234 have been lost. The limiting case where a pivot $a_{kk}^{(k)}$ is zero would give total loss of information and must be avoided completely but it is also important to control partial loss of information. The precise mathematical result upon which these remarks are based is that the LU factorization obtained is the exact factorization of a perturbed matrix A+F where the perturbations are bounded by the inequalities

$$|f_{ij}| \le (3.01)\ \varepsilon\ m_{ij} \max_k |a_{ij}^{(k)}| \tag{5.2}$$

where ε ($< 10^{-3}$) is the relative accuracy of the computation and m_{ij} is the number of non-zero products $a_{ik}^{(k)}\ a_{kj}^{(k)}$ (see (4.2)). This is a minor extension (Reid,1971) of Wilkinson's backward error analysis.

If all matrix elements are of comparable size (and fortunately it often happens that problems occur naturally like this) then we may introduce row and column interchanges with the aim of controlling the size of the largest matrix element at each stage (see §12), that is use (5.2) in the weakened form

$$|f_{ij}| \le (3.01)\ \varepsilon\ m_{ij} \max_{i,j,k} |a_{ij}^{(k)}| \ . \tag{5.3}$$

Such a bound is obviously inadequate for badly scaled matrices. The problem of scaling matrices automatically remains essentially unsolved, although Curtis and Reid (1972) report favourably on the use of Hamming's (1971) suggestion of minimizing

$$\sum_{a_{ij}\neq 0} (\log |a_{ij}| - \rho_i - c_j)^2 \tag{5.4}$$

to give the scaled matrix $\{e^{-\rho_i}\ a_{ij}\ e^{-c_j}\}$. A similar method was proposed by Fulkerson and Wolfe (1962), who minimized

$$\max_{a_{ij}\neq 0} |\log|a_{ij}| - \rho_i - c_j| \tag{5.5}$$

but Curtis and Reid found that this gives worse results than Hamming's suggestion. They also report that very poor results were sometimes obtained by such simple procedures as scaling all rows to have largest element of modulus unity, then doing likewise by columns. The Hamming procedure is not expensive to implement, normally requiring 7-10 sweeps through the matrix. Strictly speaking, in view of (5.3), we should decide what perturbation is tolerable for each matrix element and scale these rather than the matrix elements themselves.

6. Limiting fill-in

Anotherreason for using row and column interchanges is to limit the number of fill-ins, that is zero entries that become non-zero. For example if elimination is applied to the matrix of figure 6.1 then $A^{(2)}$ is totally full, but if the first and last rows and columns are exchanged then the matrix of figure 6.2 is obtained and when elimination is applied to this no new non-zeros are created.

$$\begin{bmatrix} x & x & x & x & x \\ x & x & & & \\ x & & x & & \\ x & & & x & \\ x & & & & x \end{bmatrix} \qquad \begin{bmatrix} x & & & & x \\ & x & & & x \\ & & x & & x \\ & & & x & x \\ x & x & x & x & x \end{bmatrix}$$

Figure 6.1 Figure 6.2

The most obvious case where fill-in can be avoided is when the matrix begins as triangular. More generally it may be <u>block triangular</u>, for instance of the form

$$A = \begin{pmatrix} A_{11} & & & \\ A_{21} & A_{22} & & \\ \vdots & & \ddots & \\ A_{N1} & A_{N2} & \cdots & A_{NN} \end{pmatrix} \tag{6.1}$$

where the diagonal blocks are square. The corresponding equation (4.1) can then be solved as the sequence of smaller problems

$$A_{kk}x_k = b_k - \sum_{j=1}^{k-1} A_{kj}x_j, \qquad k=1,2,\ldots,N. \tag{6.2}$$

Notice that the off-diagonal blocks appear only in a multiply operation in (6.2) and that fill-in is confined to that produced within the diagonal blocks A_{kk} when solving the sets of equations (6.2).

7. Finding a transversal

The first step in obtaining the form (6.1) is to perform interchanges that place non-zeros on the main diagonal. Such a set of non-zeros is called a <u>transversal</u> and exists whenever the matrix is non-singular. An effective algorithm is that of Hall (1956), which we describe briefly. A fuller description has been given by Duff (1976c). After k steps we have non-zeros in the first k diagonal positions. If there is a non-zero in the submatrix of rows and columns k+1 to n then a simple row and column exchange brings it to the $(k+1)^{st}$ diagonal position and the step is complete. If not, then we look for a path of the kind illustrated in figure 7.1, starting from a non-zero of column k+1 and ending at a non-zero in a row beyond the kth. Column interchanges may then be used to get this non-zero into column k+1 without destroying the diagonal structure in the leading kxk submatrix. In the example of figure 7.1 k is 7 and the exchange of columns 8 and 1 preserves the non-zero in position (1,1) and leaves us with a (shorter) path from a_{38} to a_{94}. Further exchanges between column pairs (8,3), (8,6), (8,4) give us a non-zero in position (9,8) which may be brought to the required position (8,8) by exchanging rows

(8,9). The path we want must not traverse any row or column more than once so we mark each row and column as we go along it and from each diagonal element we search its column for a non-zero in rows k+1 to n or failing this in an unmarked row. If there are none then we mark that column and go back to the previous column on our path and try there to find a non-zero in an unmarked row; failing this we try the previous column, and so on. On finding such a non-zero we use it to continue our path. In this way we must eventually find a path of the kind we want unless at some stage all non-zeros in marked columns are in marked rows. This means that there are r+1 columns which contain non-zeros in only r rows so that we have discovered that the matrix is singular. Notice that while looking for our path we look at the non-zeros in each column at most twice, once when searching for a non-zero beyond row k and a second time when searching for other non-zeros in unmarked rows. It follows that if the matrix has τ non-zeros then no more than $O(\tau)$ operations are needed to extend the zero-free diagonal by one and no more than $O(n\tau)$ operations are needed in all. Unfortunately sets of examples exist which show this $O(n\tau)$ behaviour (see Duff (1976c)), but it is extremely rare in practical problems. Duff reports that the operation count is usually a very modest multiple of τ and he prefers this algorithm to the more complicated method of Hopcroft and Karp (1973) which has an $O(\tau\sqrt{n})$ bound on operation count but executes significantly slower on all his test examples except those designed specifically to demonstrate that the Hall algorithm may involve $O(n\tau)$ operations.

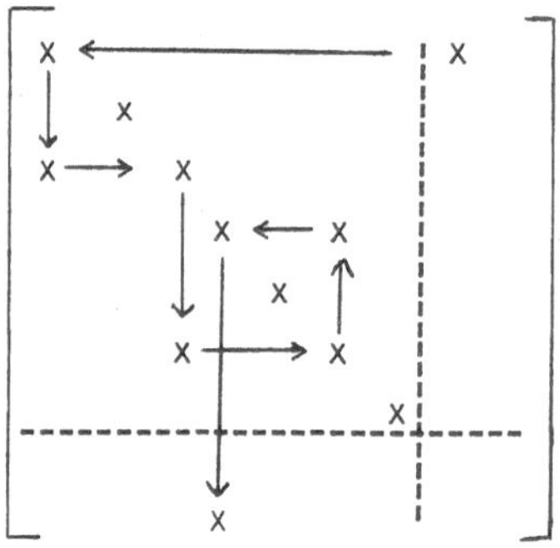

Figure 7.1

8. Finding the block triangular form: general remarks

Once we have non-zeros on the diagonal we may construct the digraph of the matrix (see §2) and use this to help in finding a symmetric permutation P^TAP which gives the required triangular form. Such a permutation corresponds to a change in the way nodes are numbered but to no other change in the graph. If we cannot find a closed path passing through all the nodes of the digraph then we must be able to divide it into two parts which are such that there is no connection from the first part to the second. Renumbering the first group of nodes 1,2,...,k and the second k+1,...,n will produce a corresponding (permuted) sparse matrix in block lower triangular form. An example is shown in figure 8.1. The same process may now be applied to each resulting block, continuing until we have broken the graph into strong components, which each contain closed paths passing through all their nodes, and cannot be enlarged by adding extra nodes without losing this property. The graph of figure 8.1 contains just two strong components and the corresponding submatrices cannot be permuted to block triangular form.

$$\begin{bmatrix} x & x & & & \\ x & x & & & \\ & x & x & x & \\ & & & x & x \\ & & x & & x \end{bmatrix}$$

Figure 8.1

In the next two sections we describe two alternative algorithms for finding all the strong components. They both depend on the original transversal, however, and so there remains some uncertainty as to whether another transversal might not have produced a better result. In fact Duff (1976b) has shown that the block triangular form is essentially unique, the only possible permutations being combinations of permutations within each block and permutations between blocks as a whole. We sketch a proof of this result. Another transversal must consist of a set of non-zeros no two of which are in the same row or column. These must all come from the diagonal blocks of the form (6.1) [because the non-zeros for the n_1 rows corresponding to A_{11} must come from the first n_1 columns, that is from A_{11} itself; once we have used the first n_1 rows we may apply the same argument to the second block and so on]. Therefore a column permutation that only permutes within the diagonal blocks can be used to get from the first form (6.1) to another form which looks exactly the same but has the other transversal on its diagonal. If we are able to get smaller diagonal blocks by applying one of our algorithms to this new form then the inverse column permutation would yield a further reduced version of (6.1), in contradiction to the assumption that it is fully reduced. Thus the block lower triangular form is essentially unique, the only possible permutations being combinations of permutations within each block and permutations between blocks as a whole. Thus the work in the block forward substitution (6.2) is fixed although the exact form of each step and the order in which the steps are performed is not necessarily fixed.

The algorithm of Tarjan (1972), described in section 10, is extremely satisfactory, since it makes modest computational and storage demands and can be coded simply. It is however much more difficult to understand than the classical algorithm of Sargent and Westerberg (1964), which is almost as efficient if coded very carefully. We therefore describe both algorithms.

9. Finding the block triangular form: the algorithm of Sargent and Westerberg

The algorithm of Sargent and Westerberg (1964) traces paths in the graph associated with A and in modifications of this graph. Starting from any node, a path is followed through the graph until a cycle is found (identified by encountering the same node twice) or a node is encountered with no edges leaving it. All nodes in a cycle must belong in the same strong component and when one is found the graph is modified by collapsing all the nodes in the cycle into a single "composite node"; edges between constituent nodes are ignored and edges entering or leaving constituent nodes from elsewhere in the graph are regarded as entering or leaving the new composite node. If a node (or composite node) is encountered with no edges leaving it then it must correspond to a strong component of the original graph having no edge to any other strong component. This therefore corresponds to the first block of the required block triangular form. The composite node and its associated edges are then deleted from the graph to leave a graph corresponding to the permuted matrix A with its first block row and column removed. The algorithm continues from the last node on the remaining path or starts from another node if the path has become empty. In this way the blocks of the required form are obtained successively.

The algorithm has the great virtue that each non-zero off-diagonal element (corresponding to an edge in the original graph) has to be inspected only once. Its disadvantage lies in the overheads associated with recording which nodes belong in each composite node and to which composite node each node belongs. In the example of figure 9.1 successive composite nodes are

$$(4,5),(3,4,5,6),(2,3,4,5,6,7),(1,2,3,4,5,6,7,8).$$

A simple scheme such as labelling each composite node with the lowest node number of a constituent node could result in almost every node being identified with n different composite nodes necessitating $O(n^2)$ relabellings. The examples of figures 9.1 to 9.4, where n can be arbitrarily large, are likely to show such a behaviour if an inferior implementation is adopted. Munro (1971a) suggested that successive amalgamations of composite nodes should be performed by relabelling only the one with least constituent nodes. He showed (Munro,1971b) that the resulting algorithm involves at most $O(n \log n) + O(\tau)$ operations and that this bound is attained by the example of figure 9.4. A more satisfactory scheme is described by Tarjan (1975). The nodes of each composite node are regarded as belonging to a tree, that is each node except one (the root node) is regarded as having another node as its father. The root node is used as a label for the

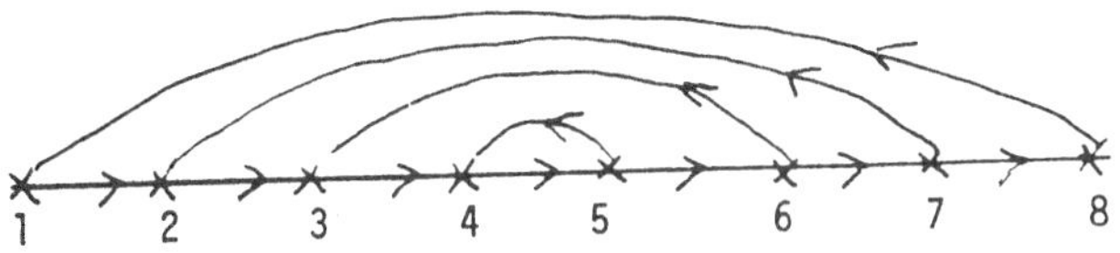

Figure 9.1 Concentric semi-circles (n= 2m, τ = 5n/2-1)

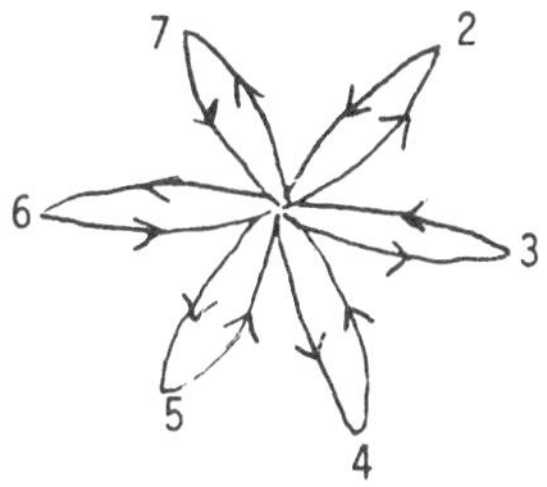

Figure 9.2 Star pattern (any n, τ = 3n-2)

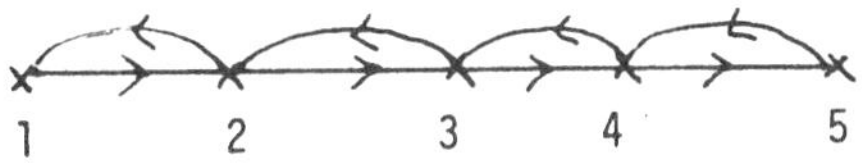

Figure 9.3 Graph of tridiagonal matrix (any n, τ = 3n-2)

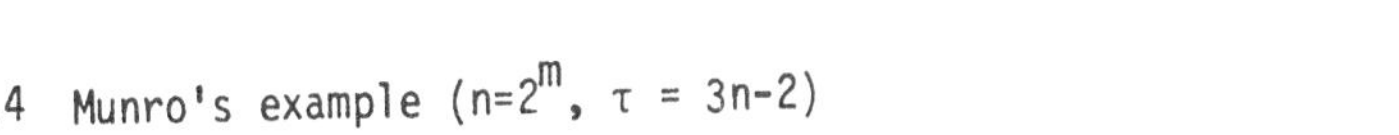

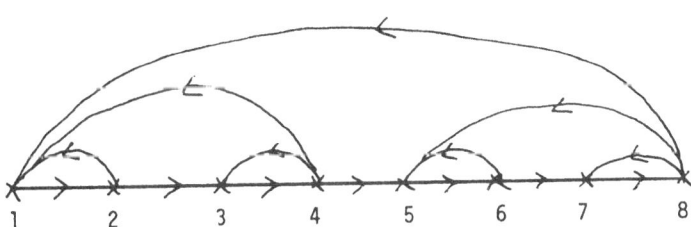

Figure 9.4 Munro's example ($n=2^m$, τ = 3n-2)

composite node. Joining together two composite nodes requires only that the root of one tree, preferably that with fewer nodes (as in Munro's algorithm), is taken to have the other root as its father. To find the composite node to which a given node belongs requires only that the path through successive fathers is followed to the root; later work is saved if, once the root is found, all nodes on this path are regarded as having the root as their father. Tarjan shows that the resulting algorithm requires O(n log*n) operations where

$$\log^*n = \min\{i: \overbrace{\log\log\log\ldots\log}^{i\text{ times}}(n) \le 1\}$$

to make the overall algorithm $O(\tau + n\log^*n)$. Notice that log*n is a function that increases with n incredibly slowly.

10. Finding the block triangular form: the algorithm of Tarjan

Tarjan (1972) proposed an algorithm similar to that of Sargent and Westerberg but with the advantage of avoiding the potentially expensive node collapsing steps. Again paths through the graph are followed, but now a stack of nodes is maintained. This stack holds those nodes which are on the current path or from which a "backtrack" has taken place and it will be shown that the required strong components can be read successively from the top of the stack. Nodes are put on the path and stack when first encountered and are removed from the path but not the stack when backtracking takes place. With each node on the stack is held a pointer, called "lowlink", to the node lowest on the stack to which a path has so far been found. Lowlink is initialized to the stack position of the node itself.

The algorithm consists of a number (possibly one) of major steps each comprised of a sequence of minor steps. A major step begins by placing on the stack and on the path any node that has not been on the stack in a previous major step. It then executes a sequence of minor steps, each of which extends the path by one more node or reduces its length by one by backtracking. The major step terminates when the path and stack become empty. A minor step begins with a search of the edges leading from the node v at the end of the path, excluding those previously searched. If an edge points to a node w whose lowlink points lower in the stack than the lowlink of v then the lowlink of v is reset to the lowlink of w. The search continues until

<u>either</u> i) an edge is found entering a node not on the stack; in this case the node is added to the stack, the path is extended by putting the node at the end of it, and the minor step is complete.

<u>or</u> ii) the list of edges leaving the node at the end of the path is exhausted; in this case lowlink is examined and

<u>either</u> a) lowlink points to the node v currently at the end of the path; in this case v is called the root of a block and v and all nodes above it on the stack consist of the next strong component (to be shown

below). They are therefore removed from the stack and from further consideration and are numbered next in the ordering array. The minor step is completed by backtracking to the node previous to v on the path unless the path, and stack, have become empty (ending the major step).

or b) lowlink points to a node lower on the stack than the node v currently at the end of the path. In this case the step is completed by backtracking to the node w prior to v on the path. Lowlink for w is reset to that for v if this results in its pointing to a node lower in the stack.

The major step ends when a minor step produces an empty stack, which must eventually happen. If all the nodes have not been ordered another major step is commenced. Eventually the algorithm terminates when all the nodes have been ordered so making it impossible to start another major step.

It remains to show that if lowlink = v in ii)a) then v and the nodes above it on the stack constitute a strong component from which there is no connection to any node outside it. Since no backtracking through v has yet occurred (because it is on the path and can only get there once), v must have been on the path when each node w above v on the stack was placed there. It follows that there is a path from v to every such node w. Also backtracking from every such node w must by now have occurred so there must be a path from every such w to a node below w on the stack; furthermore none can be to a node below v because then lowlink would now point below v. There is therefore a path from v to every node w above it on the stack and back. Furthermore there cannot be any connection to any other node so far unordered. This is the required result.

Note that, as in the Sargent and Westerberg algorithm, each edge is examined just once. Note also that each node is placed on the stack just once and is involved in just one backtracking step so the number of minor steps cannot exceed 2n.

Duff and Reid (1976) provide a Fortran subroutine to implement this algorithm. Despite the difficulty we have found in describing and justifying the algorithm, the code itself is short and simple.

11. The three stages: analyse, factorize and operate

We now assume that our system cannot be reduced to block triangular form. Often systems of this kind arise naturally (for example this is almost always true for systems arising from the solution of partial differential equations), and when the original system can be permuted to block triangular form, we obtain a sequence of such systems to solve.

A substantial amount of work may be involved in choosing suitable interchanges and following the progress of the sparsity pattern during the course of the elimination. Fortunately it often happens that we have a sequence of problems to solve all of whose matrices have the same sparsity pattern and whose elements do not differ substantially. This may occur, for example, when solving a non-linear system of equations by Newton's method or by a quasi-Newton method, the matrices A being successive Jacobians or approximations to them. Another example occurs when using an implicit method for solving a system of ordinary differential equations when the matrix elements (but not the pattern) vary from time step to time step. In such a case second and later systems may be solved far faster than the first if the same interchanges may be used again, because not only do we save the time involved in choosing interchanges but also we know where all the fill-ins take place and so we can use a simpler data structure. We will call these two stages:

1) ANALYSE: when the matrix is analysed to find the interchanges and fill-ins and Gaussian elimination is applied to the permuted matrix PAQ to produce the factorization LU as in equation (4.5), and

2) FACTORIZE: when a new matrix with the same structure is treated using information stored by ANALYSE.

We also consider a third stage:

3) OPERATE: which takes a vector b and performs the operation of calculating $A^{-1}b$ using the factorized form produced by an earlier ANALYSE or FACTORIZE.

This third stage is very useful where we have a sequence of problems with the same matrix, as we do at a single time step during the course of solution of a system of non-linear differential equations by an implicit method. An idea of the relative speeds of the stages in practice is given by the results of Curtis and Reid (1971b) shown in Table 11.1. The author estimates that code for a full matrix of order 199 would take about 16,000 msecs for FACTOR and 240 msecs for OPERATE. Even greater gains are normal in problems of larger order.

Table 11.1

Matrix order		54	57	199
Non-zeros in A		291	281	701
Non-zeros in L/U		398	360	1,542
Time in 360/75 milli-secs	ANALYSE	400	330	2,000
	FACTORIZE	60	40	260
	OPERATE	13	12	47

We conclude this section by discussing when a matrix is sparse. Firstly it depends on whether we have a "one-off" or a "many-off" problem. If we are solving a single nxn problem then we must compare ANALYSE times with comparable times for full matrices whereas if we are solving a sufficiently large number of problems with the same matrix or the same structure then it is appropriate to neglect ANALYSE times and consider only FACTORIZE and OPERATE times. Examination of timings such as those in Table 11.1 suggest that a one-off job should be not more than about 10% dense after all fill-ins for sparsity code (ANALYSE) to be more economical in computer time than full code. Comparable break-even points when the FACTORIZE and OPERATE parts of the code dominate are about 50% and 35%, respectively, for the Harwell (Curtis and Reid, 1971a) code and 100% for in-core loop-free code (see section 14) since this has execution time that cannot be improved even for a full matrix. Storage may also determine whether a sparse code is used. For example it may be possible to hold a 20% dense matrix in main store but not its expanded form. Storage break-even points of course depend on relative word-lengths of reals and integers, but they will usually be higher than the figures of 10%,50%, 35% that I gave above for computing break-even points.

12. The analyse stage: strategy

We now consider the analyse stage. The usual way of obtaining stability is to choose a parameter u in the range $0<u\leq 1$ and then to use interchanges to ensure that one of the conditions

$$\left.\begin{aligned} |a_{kk}^{(k)}| &\geq u \max_{k\leq i\leq n} |a_{ik}^{(k)}| \\ |a_{kk}^{(k)}| &\geq u \max_{k\leq j\leq n} |a_{kj}^{(k)}| \end{aligned}\right\} \qquad (12.1)$$

holds so that we may deduce from equation (4.2) the inequality

$$|a_{ij}^{(k+1)}| \leq (1+u^{-1}) \max_{i,j} |a_{ij}^{(k)}| \qquad (12.2)$$

so the growth in size of matrix elements is limited at each stage of the elimination to the factor $1+u^{-1}$. Ordinary partial pivoting corresponds to the choice u=1 which does not allow enough freedom for good exploitation of sparsity. Curtis and Reid (1971b) found that the choice of u is not very critical and recommended $u=\frac{1}{4}$ on the grounds of numerical experiments (and this value was used for the Table 11.1 results). Tomlin (1972) recommended the rather low figure u=1/100 for linear programming matrices. The reason for such a small value being tolerable is that the elements in a particular column j only change if it contains a non-zero a_{kj} in the pivot row; it follows that the overall growth in column j is limited by the factor

$$(1 + u^{-1})^{p_j - 1} \tag{12.3}$$

if p_j is the total number of non-zeros, including fill-ins, in column j. Particularly in the linear programming cases that Tomlin was considering, the number p_j may be quite small (e.g. 3,4 or 5) so that the factor (12.3) need not be enormous, in stark contrast to the full case. It is likely, however, to be larger than we are prepared to tolerate for the actual growth. It is not a good estimator for this actual growth which is likely to be much smaller, so for safety we need to monitor the size of the elements $a_{ij}^{(k)}$ or find a better estimator. Such an estimator was suggested by Erisman and Reid (1974). If Gaussian elimination produces the factorization

$$PAQ = L\ U \tag{12.4}$$

where P and Q are permutation matrices, L is unit lower-triangular and U is upper-triangular then

$$a_{ij}^{(k)} = a_{ij} - \sum_{m=1}^{k} \ell_{im}\, u_{mj}; \quad k<i\leq n \quad \text{and} \quad k<j\leq n \tag{12.5}$$

so that Hölder's inequality with $1/p + 1/q = 1$ gives the bound

$$|a_{ij}^{(k)}| \leq |a_{ij}| + \|(\ell_{i1},\ell_{i2},\ldots,\ell_{ik})\|_p\ \|(u_{1j},u_{2j},\ldots,u_{kj})\|_q\ . \tag{12.6}$$

If $p=1$ and $q=\infty$ or vice-versa then these norms may be accumulated in 2n variables and applied to the non-zeros in the pivotal row and column to give a bound for their size over the earlier history of the elimination. In this way growth monitoring is taken out of the inner loop of the program. Alternatively inequality (12.6) may be weakened to

$$|a_{ij}^{(k)}| \leq \max_{i,j} |a_{ij}| + \max_{i} \|(\ell_{i1},\ldots,\ell_{i,i-1})\|_p \max_{j} \|(u_{1j},\ldots,u_{j-1,j}\|_q \tag{12.7}$$

which is an a posteriori bound involving the norms of the rows of the sub-diagonal part of L and the norms of the columns of the superdiagonal part of U. If the growth is found in one of these ways to be unsatisfactory then a new ANALYSE with a larger value for u will be needed.

It should be noted that the generalisation of the factorization (12.4) to rectangular and rank-deficient cases is straightforward. We merely terminate elimination when no more non-zero pivots are available.

Next we consider sparsity criteria for the choice of interchanges. The criterion of Markowitz (1957) has proved to be satisfactory over a very wide range of problems. It is to find the non-zero in $A^{(k)}$ which satisfies our stability criterion (12.1) and has the smallest product of number of other non-zeros in its row and number of other non-zeros in its column and then use interchanges to bring this non-zero to the pivotal position (k,k). This gives a local minimization of the multiplication count for the kth stage of the elimination (local in the sense that no account is taken of the possibility that a different set of earlier interchanges might have reduced further the number of multiplications involved in the kth stage). It might be thought that it would be better to minimise the number of fill-ins involved at the kth stage, but such a strategy is impossible to code as efficiently and Duff and Reid (1974) found that the gain in sparsity is never more than slight and there can even sometimes be a loss. A comparison on the three matrices used for Table 11.1 is shown in Table 11.2.

Table 11.2

Matrix order		54	57	199
Non-zeros in A		291	281	701
Non-zeros in L/U	Markowitz	381	315	1,387
	Min. of fill-in	361	291	1,372
ANALYSE time (m.sec on 370/165)	Markowitz	140	110	900
	Min. of fill-in	800	500	6,400

The differences in the Markowitz results from those of Table 11.1 are caused by the use of a faster computer and a small value of u (actually u was zero for these results because they were generated for a comparison of dependency on the sparsity structure).

A disadvantage of the Markowitz scheme is that it needs the sparsity pattern of the whole matrix at each stage and so we need a data structure that can hold it and update it as the elimination proceeds. An alternative is to order the columns before commencement of the elimination and process them one by one, that is perform all the operations for the jth column before doing any for the $(j+1)^{st}$ column. There is no difficulty about this reordering of the operations provided the multipliers $a_{ik}^{(k)}[a_{kk}^{(k)}]^{-1}$ are stored. The actual arithmetic is identical. A simpler data structure is possible because the fill-ins in each column happen together. A number of strategies for choosing the original column order have been proposed but Duff and Reid (1974) have found experimentally that the simplest possible, namely to order them by increasing numbers of non-zeros, is as satisfactory an algorithm as any. For the row interchange they recommend bringing to the (k,k) position the non-zero in the kth column which satisfies the stability criterion (12.1) and whose row had least non-zeros in the original matrix. Of course we cannot use the number of non-zeros in the rows of $A^{(k)}$ because we are not updating the sparsity pattern. It is interesting that if we try taking the number of non-zeros in the part of A that corresponds to $A^{(k)}$ we can get significantly worse results, presumably because the original number of non-zeros in the row of A is a better estimate of the number of non-zeros in the row of $A^{(k)}$ on account of the cancellation effect between eliminations and fill-ins. This effect is illustrated in Table 11.3 in the case of the matrix of order 199 and we also illustrate in this table that a priori column ordering gives significantly more fill-in than Markowitz.

Table 11.3

		54	57	199
Matrix order		54	57	199
Non-zeros in A		291	281	701
Non-zeros in L/U	Markowitz	381	315	1,387
	A priori using A	475	408	2,767
	A priori using $A^{(k)}$	461	427	5,991

13. The analyse stage: implementation

In this section we will consider the efficient implementation of the various algorithms that we have already discussed. This depends critically on the proper choice of data structure and unfortunately no one can be regarded as best in all cases.

For Markowitz' strategy we have to follow all the fill-ins, which suggests the linked list (with row and column links) or the Gustavson scheme (rows held as file of packed vectors and pattern of the columns held similarly). Both structures were described in section 3 and the way they handle eliminations and fills was explained. Also explained was how a full vector of workspace may be used to facilitate the basic operation of adding a multiple of one row to another and both structures permit ready access to the non-zeros in a row. We also need to know which rows are active at a single pivotal step. With the linked list we follow links in the pivotal column and with Gustavson's scheme we look at the sparsity pattern of the pivotal column. It is also important to avoid an expensive search for the pivot. A satisfactory strategy is to search the rows and columns in order of increasing numbers of non-zeros, rejecting any non-zeros that do not satisfy the relative pivot tolerance (12.1) and maintaining a record of the one with best Markowitz cost so far found. If the row or column currently being searched has k non-zeros and the best non-zero so far found has Markowitz cost not greater than $(k-1)^2$ then we can immediately terminate the search and use this non-zero as pivot. In order to be able to search the rows and columns in order of increasing numbers of non-zeros Westerberg (private communication) suggested maintaining linked lists (with forward and backward pointers) for rows and columns with equal numbers of non-zeros. This has been implemented by Reid (1976) and is more efficient than the earlier code of Curtis and Reid (1971a) which maintains ordered lists of rows and columns.

A simpler structure is possible with a priori column ordering and it is usual to use a file containing the columns as packed vectors. If all the operations on a column are performed together then there is no difficulty about handling the fill-ins since they all occur at this time. This is the structure favoured by linear programmers for their very big matrices which cannot be held in main storage. They make a separate backing-store (e.g. disk) file for the factorized matrix. A column is brought down from the original matrix file, operated on in main storage (and this involves reading appropriate columns of the file containing the factorized matrix) and the final column (probably longer because of fill-ins) is written to the factorized matrix file.

14. Implementation of the factor and operate phases

In section 11 we mentioned that it often happens that solutions are required to a sequence of problems with matrices having the same elements or the same structure so that it is worthwhile to arrange that FACTORIZE and OPERATE execute rapidly. Fastest execution is obtained by generating loop-free code in the ANALYSE stage. For example, if the 3x3 matrix whose structure is shown in figure 14.1 is stored in

x		x		1		4
x	x	x		5	3	2
	x	x			7	6

Figure 14.1 Figure 14.2

an array A in the position shown in Figure 14.2 then the following code would suffice for FACTORIZE:

```
A(5) = A(5)/A(1)
A(2) = A(2) - A(5)*A(4)
A(7) = A(7)/A(3)
A(6) = A(6) - A(7)*A(2).
```

This code is written in Fortran but the usual practice is to generate the equivalent machine code directly since there is no need to have a compiler for such simple code. The machine code consists of a long sequence of floating-point instructions without any loops. The only snag is that for large problems the amount of code can be huge. For example the matrix arising from the 5-point finite-difference approximation to Laplace's equation on a 30x31 grid would require about a million bytes for FACTORIZE and a third of a million bytes for OPERATE. At some cost in execution time (an increase to about $2\frac{1}{2}$ to 3 times as long) we may replace the floating-point instructions by integer arrays in which the required operations are held in encoded form and interpreted at run time. This saves some storage particularly in the complex case, but the author prefers to hold the matrix as a file of row vectors stored contiguously so that the only storage needed in addition to that for the non-zeros themselves is that for column numbers and the positions of row starts. This can be done efficiently, without any O(n) searches, if all the operations that involve any one row are performed together. Curtis and Reid (1971a) use a linked list for ANALYSE and then reorder the matrix so that PAQ is held in a file of column vectors with zeros held explicitly where fill-ins are known to occur. This structure is also suitable for OPERATE.

Additional facilities can be built into the loop-free code approach. Particularly important is the idea of variability typing (Hachtel,1972). We have already indicated how to exploit the feature that the matrix structure remains fixed while the elements vary but have not considered the possibility that some of the elements may be constants or may be constant temporarily because of a program structure consisting of nested loops. For example, if we are designing a circuit we may have an outer loop in which certain design parameters are adjusted, an inner loop in which time steps through the solution of a differential equation are made and an innermost loop involving successive linearizations of the non-linear equations to be solved at a single time step. At each level in such a program we would like to do as little work as possible. We therefore give each non-zero a variability type which is an integer giving the greatest depth within the nested loop structure at which it is changed, so that it has value 0 for constants, 1 for variables changed in the outermost loop, 2 for those changed in the next loop and so on. When an elimination step (4.2) is performed the variability type of the result is taken to be the greatest variability type of any of the elements in the right-hand side expression. During a subsequent FACTORIZE this operation may be omitted in a loop whose depth is greater than this variability type for nothing will have changed since the last FACTORIZE and so the old result will still be correct. Our pivotal strategy during ANALYSE must now aim to increase variability types as little as possible so that as many as possible of the steps may be omitted in the inner loops. For example we might minimize a weighted sum of Markowitz cost and variability type. This of course leads to a slowing down of the ANALYSE code but the gains in FACTORIZE and OPERATE can be enormous.

It should be noted that FACTORIZE uses a predetermined pivot sequence which may lead to instability. It is therefore important to monitor the size of the non-zeros either directly or through the technique of Erisman and Reid (1974), described in §12. If the growth proves unsatisfactory then normally a new ANALYSE is necessary, but Stewart (1974) suggests simply changing very small pivots and then correcting later with the Sherman-Morrison formula. Stewart's suggestion leads to an overhead in each subsequent OPERATE so should be used only when ANALYSE time is important.

Our final remarks in this section pertain to exploitation of sparsity in the vector b. If $b_k^{(k)}=0$ then the forward substitution operations (4.3) become null and may be omitted; sometimes this happens very frequently to give a very substantial saving in operations. It is tempting to hold b in packed storage since the dominant operation count may now come from testing whether $b_k^{(k)}$ is zero (this has often been our experience with linear programming problems (Reid,1976)). However the code would have to be quite complicated, making the procedure worthwhile only if the vector remains extremely sparse.

We may similarly exploit zero components x_j in the back-substitution (4.4) if all the operations involving each x_j are performed together rather than in the order suggested by (4.4), but this is less likely to be worthwhile because there will have been too many opportunities for fill-ins.

15. Symmetric positive-definite cases

In this and the next section we suppose that A is symmetric and that we wish to exploit its symmetry to save both storage and computation. It is clear from the formula (4.2) that if the matrix $A^{(k)} = \{a_{ij}^{(k)}, i\geq k, j\geq k\}$ is symmetric then so is the matrix $A^{(k+1)} = \{a_{ij}^{(k+1)}, i\geq k+1, j\geq k+1\}$. It follows that we need compute and store only the upper (or lower) triangular parts of these matrices. Of course it is important that we do not destroy the symmetry by performing an unsymmetric interchange. Fortunately it frequently happens that the matrix A is positive definite ($x^T Ax>0$ unless $x=0$), for which case Wilkinson (1961, section 6) has proved the inequality

$$\max_{i,j} |a_{ij}^{(k+1)}| \leq \max_{i,j} |a_{ij}^{(k)}| \qquad (15.1)$$

so that growth in the size of the matrix elements cannot occur. We are therefore free to choose symmetric interchanges on sparsity grounds alone. We may in fact perform our ANALYSE phase on the sparsity pattern alone, saving the storage required for the reals. Prompted by this success, early codes (and even some developed recently) have analysed the sparsity pattern of unsymmetric matrices without reference to the reals, but this is clearly unsatisfactory from a stability point of view unless there are physical grounds for, say, diagonal pivoting always being satisfactory.

The most widely-used and successful general purpose sparsity pivoting criterion is that of choosing the diagonal non-zero with least other non-zeros in its row. Notice that this is a possible Markowitz choice since there cannot be an off-diagonal element with less non-zeros. We refer to this strategy by Markowitz' name. It has been implemented by Reid (1972).

16. Symmetric non-definite cases

Wilkinson's result (15.1) is not available in the non-definite case and we have to return to processing reals in the analyse phase if we are not to risk instability. We may use the same relative pivot tolerance (12.1) as in the unsymmetric case and take for pivot the diagonal element that satisfies this and has least non-zeros in its row. However such an element may not exist (consider $\begin{bmatrix} 0 & 1 \\ 1 & 0 \end{bmatrix}$, for instance). The most satisfactory solution is to follow Bunch (1974) in using pivots $a_{kk}^{(k)}$

which may be 2x2 as well as 1x1 matrices. We need a "Markowitz" cost for 2x2 pivots. Duff and Reid (1975) suggested that for each potential 2x2 pivot (that is in association with each non-zero off-diagonal element) we take as sparsity cost the square of the number of non-zeros in the two potential pivot rows that are not in the potential pivot itself. This is a readily evaluated upper bound for the possible number of fill-ins, so generalising one interpretation of the Markowitz cost. We choose a 2x2 pivot (which results in two eliminations being performed together) if one is available with Markowitz cost less than double that of the best 1x1 pivot. For stability the requirement

$$\max_{\substack{k+2\le i\le n \\ k\le j\le k+1}} |a_{ij}^{(k)}| \; \| [a_{kk}^{(k)}]^{-1} \|_\infty \le u^{-1} \tag{16.1}$$

would appear appropriate. In this case u should be restricted to the range $0\le u\le \frac{1}{2}$, for then it is easily shown that one of the diagonal elements is suitable as a 1x1 pivot or the 2x2 pivot defined by the largest off-diagonal element is suitable.

17. Least squares: orthogonal reduction

In the next five sections we consider the solution of a sparse set of overdetermined linear equations

$$Ax = b \tag{17.1}$$

where A is an mxn matrix with m>n. We solve this problem in the sense of least squares, that is look for the least value of the quantity

$$S(x) = (b-Ax)^T (b-Ax). \tag{17.2}$$

It is readily verified that if equation (17.1) is multiplied by an orthogonal matrix Q then the objective function (17.2) is unchanged. If we can find an orthogonal matrix Q such that QA has the form

$$QA = \begin{bmatrix} U \\ 0 \end{bmatrix} \tag{17.3}$$

where U is upper triangular we will have an equivalent problem

$$\begin{bmatrix} U \\ 0 \end{bmatrix} x = Qb = \begin{bmatrix} w \\ z \end{bmatrix} \tag{17.4}$$

whose solution is obviously obtained by solving the first n equations Ux=w exactly, since nothing can be done with the rest.

The matrix Q can be built from a sequence if elementary plane (Givens) rotations that create zeros one at a time just like the successive steps of Gaussian elimination. Gentleman (1973) has shown that these Givens transformations can be applied with about half as many arithmetic operations as classically and without calculating any square roots or trigonometric functions by using the transformed sets

$$D^{(k)}A^{(k)}x = D^{(k)}b \tag{17.5}$$

where $D^{(k)}$ is a diagonal matrix whose square is held. Furthermore the number of multiplications is approximately halved from the number required by the traditional Givens matrices (though in practice execution times are not decreased by nearly so much). The operations now become those of adding a multiple of each of the two active rows to the other. It is apparent that the fill-in must be much more severe than that for Gaussian elimination since both rows may fill.

The alternative of using Householder matrices of the form $(I-2ww^T)$ to reduce all the sub-diagonal elements of the pivotal column to zero at once is even worse because here all the rows with non-zeros in the pivotal column take the sparsity pattern of their union.

Duff (1974) has experimented with pivotal strategies in connection with the Givens' orthogonal reduction and recommends that the column with least non-zeros be chosen as pivotal and the pivot be chosen as that non-zero which has least other non-zeros in its row and that the non-zeros be eliminated in order of increasing numbers of non-zeros in the rows.

18. Least squares: normal equations

Another approach is through the formation and solution of the normal equations

$$A^TAx = A^Tb \tag{18.1}$$

which may be obtained by multiplying equation (17.4) by $[U^T\ 0] = A^TQ^T$. This is less satisfactory from a numerical point of view. In fact the 2-norm condition number of A^TA is exactly the square of that of U. The disadvantage is particularly marked when the elements in some of the rows of A are much larger than the rest since row scaling cannot be applied without altering the least squares objective function. Operation counts obtained by forming A^TA and then symmetrically factorizing it (see §15) are quite favourable, so the method must be regarded as a possibility for well-conditioned problems. It is apparent from the way we constructed the normal equations that U is the transpose of the Choleskii factor L of A^TA. The various ways of calculating it, however, are very different both from the sparsity and accuracy points of view.

The actual formation of the normal matrix $B=A^TA$ is surprisingly difficult in the sparse case. The usual inner-product formula

$$b_{ij} = \sum_k a_{ki}\, a_{kj} \tag{18.2}$$

is inefficient because it demands a scan of the non-zeros in the ith (or j^{th}) column of A and many of the corresponding components in the jth (or i^{th}) column may be zero; indeed we may perform a column scan only to find that b_{ij} is zero. If n is sufficiently small for full storage to be appropriate for the nxn matrix B then the outer-product formula

$$B = \sum_i a_{i.} a_{i.}^T \quad , \tag{18.3}$$

where $a_{i.}^T = (a_{i1}, a_{i2}, \ldots, a_{in})$ is the ith row of A, is appropriate and requires only that we take products of all pairs of non-zeros in each row. Where B is to be held as a sparse matrix we may still use (18.3) if a data structure that permits fill-ins is in operation.

19. Least squares: method of Peters and Wilkinson

Another possibility, proposed by Peters and Wilkinson (1970), involves the direct application of Gaussian elimination to yield the factorization

$$A = LU \tag{19.1}$$

where L us a unit lower trapezoidal mxn matrix and U is an nxn upper triangular matrix (not the same as in equation (17.3)). They then change variables, solving the problem

$$Ly = b \tag{19.2}$$

by the normal equations method and finally calculating x by solving the equation

$$Ux = y \ . \tag{19.3}$$

We may treat the rectangular matrix A exactly as we treated a square matrix earlier. With a relative pivot tolerance there is every hope that L will be well conditioned so that using the normal equation approach to solve (19.2) is likely to be satisfactory.

20. Least squares: augmented equations

Hachtel (private communication) has suggested applying sparsity techniques directly to the (m+n) x (m+n) system

$$\begin{bmatrix} I & A \\ A^T & 0 \end{bmatrix} \begin{bmatrix} r \\ x \end{bmatrix} = \begin{bmatrix} b \\ 0 \end{bmatrix} \ . \tag{20.1}$$

We can see that this gives us the solution we want by performing a block Gaussian elimination to yield the equations

$$\begin{bmatrix} I & A \\ 0 & -A^TA \end{bmatrix} \begin{bmatrix} r \\ x \end{bmatrix} = \begin{bmatrix} b \\ -A^Tb \end{bmatrix} \ , \tag{20.2}$$

the second block of which is just the normal equations. We see from the first block of (20.1) or (20.2) that r is the residual vector b-Ax. Solving (20.1) by pivoting down the diagonal amounts exactly to forming and solving the normal equations, but an alternative is to regard (20.1) simply as a sparse set of equations. The matrix is symmetric but is non-definite, so we may treat it using a mixture of 1x1 and 2x2 pivots, as described in §16. Interestingly, however, there are some advantages in ignoring the symmetry, for we may then use a pivot choice that exploits the zeros on the right-hand side and the fact that the back-substitution may be terminated once x has been calculated.

21. Least squares: comparison of methods

Duff and Reid (1975) compared these algorithms experimentally. They found that the best from the point of view of storage and operation count was always either the use of the normal equations or Hachtel's augmented matrix. Results with the orthogonalization methods, particularly with Householder transformations, were normally significantly inferior. The Peters and Wilkinson algorithm did not behave so badly and Duff and Reid therefore recommend its use when stability is important. Otherwise they on the whole prefer the symmetric variant of Hachtel's scheme.

References

Bunch, J.R. (1974). Partial pivoting strategies for symmetric matrices. SIAM J. Numer. Anal. 11, 521-528.

Curtis, A.R. and Reid, J.K. (1971a). Fortran subroutines for the solution of sparse sets of linear equations. AERE Report R.6844. HMSO, London.

Curtis, A.R. and Reid, J.K. (1971b). The solution of large sparse unsymmetric systems of linear equations. J. Inst. Math. Appl. 8, 344-353.

Curtis, A.R. and Reid, J.K. (1972). On the automatic scaling of matrices for Gaussian elimination. J. Inst. Math. Appl. 10, 118-124.

Duff, I.S. (1974). Pivot selection and row ordering in Givens reduction on sparse matrices. Computing 13, 239-248.

Duff, I.S. (1976a). A survey of sparse matrix research. Harwell report CSS 28. To appear in Proc. I.E.E.E.

Duff, I.S. (1976b). On permutations to block triangular form. Harwell report CSS 27. To appear in J. Inst. Math. Appl.

Duff, I.S. (1976c). On algorithms for obtaining a maximum transversal. To appear.

Duff, I.S. and Reid, J.K. (1974). A comparison of sparsity orderings for obtaining a pivotal sequence in Gaussian elimination. J. Inst. Math. Appl. 14, 281-291.

Duff, I.S. and Reid, J.K. (1975). A comparison of some methods for the solution of sparse overdetermined systems of linear equations. Harwell report CSS 12. J. Inst. Math. Appl. 17, 267-280.

Duff, I.S. and Reid, J.K. (1976). An implementation of Tarjan's algorithm for the block triangularization of a matrix. Harwell report CSS 29. To appear in ACM Transactions on Mathematical Software.

Erisman, A.M. and Reid, J.K. (1974). Monitoring the stability of the triangular factorization of a sparse matrix. Numer. Math. 22, 183-186.

Fulkerson, D.R. and Wolfe, P. (1962). An algorithm for scaling matrices. SIAM Rev. 4, 142-146.

Gentleman, W.M. (1973). Least squares computations by Givens transformations without square roots. J. Inst. Math. Appl. 12, 329-336.

Gustavson, F.G. (1972). Some basic techniques for solving sparse systems of linear equations. In Rose and Willoughby (1972), 41-52.

Hachtel, G.D. (1972). Vector and matrix variability type in sparse matrix algorithms. In Rose and Willoughby (1972), 53-64.

Hall, M. (1956). An algorithm for distinct representatives. Amer. Math. Monthly 63, 716-717.

Hamming, R.W. (1971). Introduction to applied numerical analysis. McGraw-Hill, New York.

Hopcroft, J.E. and Karp, R.M. (1973). An $n^{5/2}$ algorithm for maximum matchings in bipartite graphs. SIAM J. Comput. 2, 225-231.

Markowitz, H.M. (1957). The elimination form of the inverse and its application to linear programming. Management Science 3, 255-269.

Munro, I. (1971a). Efficient determination of the transitive closure of a directed graph. Information Processing Lett., 1, 55-58.

Munro, I. (1971b). Some results in the study of algorithms. Ph.D. Thesis, Dept. of Comp. Sci., Toronto. Report # 32.

Peters, G. and Wilkinson, J.H. (1970). The least squares problem and pseudo-inverses. Comput. J. 13, 309-316.

Reid, J.K. (1971). A note on the stability of Gaussian elimination. J. Inst. Math. Appl. 8, 374-375.

Reid, J.K. (1972). Two Fortran subroutines for direct solution of linear equations whose matrix is sparse, symmetric and positive definite. AERE Report R.7119. HMSO, London.

Reid, J.K. (1976). Fortran subroutines for handling sparse linear programming bases. AERE Report R.8269. HMSO, London.

Rose, D.J. and Willoughby, R.A. (Ed.)(1972). Sparse matrices and their applications. Proc. of Conf. at IBM Research Center, N.Y. Sept. 9th-10th 1971. Plenum Press, New York.

Sargent, R.W.H. and Westerberg, A.W. (1964). "Speed-up" in chemical engineering design. Trans. Inst. Chem. Engrgs. 42, 190-197.

Stewart, G.W. (1974). Modifying pivot elements in Gaussian elimination. Math. Comp. 28, 537-542.

Tarjan, R.E. (1972). Depth first search and linear graph algorithms. SIAM J. Comput. 1, 146-160.

Tarjan, R.E. (1975). Efficiency of a good but not linear set union algorithm. J. ACM 22, 215-225.

Tomlin, J.A. (1972). Pivoting for size and sparsity in linear programming inversion routines. J. Inst. Math. Appl. 10, 289-295.

Wilkinson, J.H. (1961). Error analysis of direct methods of matrix inversion. J. ACM 8, 281-330.

Zlatev, Z. and Thomsen, P.G. (1976). ST - A FORTRAN IV SUBROUTINE for the solution of large systems of linear algebraic equations with real coefficients by use of sparse technique. Report 76-05, Numerisk Institut, Danmarks Tekniske Højskole.

COMPUTATION OF EIGENVALUES AND EIGENVECTORS

Axel Ruhe

Institute of Mathematics and Statistics
University of Umeå
S 901 87 Umeå, Sweden

CONTENTS

Research supported in part by the Swedish Natural Science Research Council.

1. Sources of eigenvalue problems

It is a well-known fact that a rich variety of problems in the Natural, Engineering and Social sciences might to great advantage be solved by matrix methods with linear systems of equations, linear least squares problems and algebraic eigenvalue problems as the most common formulations. In order to motivate our interest in eigenvalue computations, we start by giving a few examples of large sparse eigenvalue problems that will give us some idea of the questions that are asked and where the difficulties are.

1. Vibrations of a mechanical structure. The classical example of an eigenvalue problem is given by a mechanical system oscillating around a point of equilibrium.

Consider a mass point of mass m, fastened elastically at the origin with stiffness (spring constant) k. Then balancing static and kinetic energy gives us the relation

$$m\ddot{x}+kx = 0 .$$

($\ddot{x}$ denotes the second time derivative, acceleration). As is well known, this equation has periodic solutions

$$x = x_o \sin \omega t,$$

with the eigenvalue ω^2 determined as $\omega^2 = k/m$.

A large interconnected mechanical system can likewise be described by a mass matrix M and a stiffness matrix K where x is now a vector and the equation is

$$M\ddot{x}+Kx = 0.$$

Assuming all components oscillate at the same frequency

$$x = y \sin \omega t$$

we get

$$-M\omega^2 y + Ky = 0,$$

a (generalized) eigenvalue problem

$$(K-\lambda M)y = 0.$$

In all realistic cases K and M are symmetric. Furthermore, M is positive definite and in simple cases diagonal. K is positive semidefinite.

The same description is applicable also to systems that are continuous in space, such as membranes or elastic bodies. They are reduced by methods of finite differences or finite elements, giving stiffness and mass matrices of a very regular shape.

If we, for example, apply the method of finite elements with first-order rectangular elements to an L-shaped membrane we get the matrices K and M given by fig. 1.1.

Another related eigenvalue problem occurs in buckling analysis of a structure. Now we get

$$(K-\lambda K_G)y = 0$$

where K is the above mentioned (small deflection) stiffness matrix and K_G is the geometric stiffness matrix. In this case K_G is in general indefinite, so the reversed problem with $\xi = 1/\lambda$ as solution is solved in practice.

2. Wave functions, the Schrödinger equation. In quantum mechanics we describe phenomenae by means of wave functions, and a most fundamental law describing them is the Schrödinger equation,

$$-\Delta\varphi+P\varphi-\lambda\varphi = 0,$$

here in time-independent form. Δ denotes the Laplacian operator, φ the wave function and P the potential function. Making a finite difference discretization, we get from $-\Delta$ a well-known positive definite symmetric matrix and from P a diagonal matrix. Normally P assumes large negative values in holes, in interesting cases so deep that the combined matrix is no longer positive definite and we have bound states. In one case we have considered we work over the plane with the holes forming a regular array, all the same size except one or a few (so-called dislocations). Since we need at least 10 points between two holes, that makes the order n of the matrix about 100 times the number of holes. A 20x20 array of holes thus gives us a matrix of the order 40 000, a clearly unmanageable size.

Chemical configuration interaction computations also use the Schrödinger equation, now for the atoms of a set of molecules. Often other methods than finite differences are used, mainly expansions in elementary wave functions. This gives rise to filled or banded matrices of very high order. See e.g. the computations reported by Shavitt et al.[37].

Fig.1.1 Membrane eigenvalue problem
Finite element approximation.

$Ku = \mu Mu$

For each point:
$$\begin{bmatrix} 1 & 1 & 1 \\ 1 & -8 & 1 \\ 1 & 1 & 1 \end{bmatrix} \qquad \begin{bmatrix} 1 & 4 & 1 \\ 4 & 16 & 4 \\ 1 & 4 & 1 \end{bmatrix}$$

Grid:

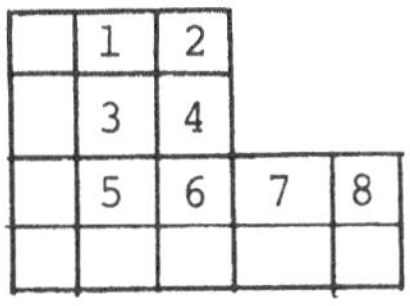

$$\tilde{K} = \begin{bmatrix} -8 & 1 & 1 & 1 & 0 & 0 & 0 & 0 \\ 1 & -8 & 1 & 1 & 0 & 0 & 0 & 0 \\ 1 & 1 & -8 & 1 & 1 & 1 & 0 & 0 \\ 1 & 1 & 1 & -8 & 1 & 1 & 1 & 0 \\ 0 & 0 & 1 & 1 & -8 & 1 & 0 & 0 \\ 0 & 0 & 1 & 1 & 1 & -8 & 1 & 0 \\ 0 & 0 & 0 & 1 & 0 & 1 & -8 & 1 \\ 0 & 0 & 0 & 0 & 0 & 0 & 1 & -8 \end{bmatrix} \qquad \tilde{M} = \begin{bmatrix} 16 & 4 & 4 & 1 & 0 & 0 & 0 & 0 \\ 4 & 16 & 1 & 4 & 0 & 0 & 0 & 0 \\ 4 & 1 & 16 & 4 & 4 & 1 & 0 & 0 \\ 1 & 4 & 4 & 16 & 1 & 4 & 1 & 0 \\ 0 & 0 & 4 & 1 & 16 & 4 & 0 & 0 \\ 0 & 0 & 1 & 4 & 4 & 16 & 4 & 0 \\ 0 & 0 & 0 & 1 & 0 & 4 & 16 & 4 \\ 0 & 0 & 0 & 0 & 0 & 0 & 4 & 16 \end{bmatrix}$$

Now $K = -\frac{1}{3h}\tilde{K}$, $M = \frac{h}{36}\tilde{M}$, and we note that

1) K and M are symmetric positive definite.
2) K and M have the same sparsity pattern.

It can be proved that M is well-conditioned with respect to inversion in all realistic cases.

3. Hydrodynamics

A spectacular example is provided by Cline, Golub and Platzman [8]. They solve the tidal equations for a finite element approximation of the Atlantic and Indian oceans.

We get the differential equation eigenproblem

$$L\underset{\sim}{a} = \sigma \underset{\sim}{a}$$

where L is a skew-hermitian matrix obtained from element approximation of the operator (in each point),

$$a \equiv \begin{pmatrix} \xi \\ u \\ v \end{pmatrix} \quad \text{and} \quad L \equiv i \begin{pmatrix} 0 & \frac{\partial}{\partial x}h & \frac{\partial}{\partial y}h \\ g\,\frac{\partial}{\partial x} & 0 & -f \\ g\,\frac{\partial}{\partial y} & f & 0 \end{pmatrix}$$

$$\begin{cases} f & \text{Coriolis parameter (dependent on Earth's rotation)} \\ g & \text{gravity} \\ h & \text{depth} \end{cases}$$

$$\begin{cases} \xi & \text{surface elevation} \\ u & \text{east velocity} \\ v & \text{north velocity} \end{cases}$$

An interesting fact is that this problem (in the example giving a matrix of order 1919) has several small eigenvalues,corresponding to slow rotations, and several large. The interesting eigenvalues are those in the middle that correspond to waves with approximately the same time constants as the tide (8-100 hours). Results for two interesting eigenvectors are shown in fig.1.2.

4. Graph partitioning

Consider a graph consisting of a set of vertices V and a set of edges E connecting some of the vertices. A practical example is an electrical network of transistors and other components (See Cullum et al. [11]). A graph is described by its incidence matrix, a symmetric matrix A with each row corresponding to one vertex and $a_{ij}=1$ whenever there is an edge from vertex i to vertex j. Often the diagonal elements are given

Fig. 1.2 Tidal equations of the Atlantic and Indian oceans. (From Cline, Golub and Platzman ref. 8).*

Fig. 1. The Atlantic and Indian Oceans resolved on a grid of 6° Mercator squares. Bathymetry (km) is drawn from grid-square averages. Dotted segments of the boundary show locations of ports.

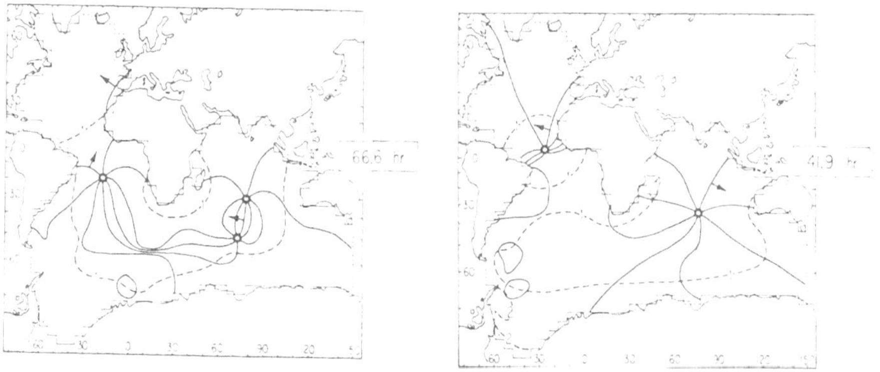

Fig. 2. Phase (solid lines) and amplitude (broken lines) of surface elevation of the first two gravity modes of the Atlantic/Indian Oceans. Upper: fundamental mode; lower: quarter-wave of the North Atlantic. Phase isogons are at intervals of 45° and an arrow in the direction of propagation is attached to the line of zero phase. (The arbitrary additive constant in the phase distribution is fixed by assigning zero phase to the grid point nearest to Lisbon.) Amplitude is normalized to an rms of 1 for the whole domain and co-amplitude lines are drawn at intervals of 1.

values so that the row sums are zero.

In some cases the eigenvalues of the incidence matrix are of interest. (See e.g. Cvetkovic [12] or Bussemaker et al. [7]). If we intend to divide the graph into q equally sized subgraphs with as few edges connecting different subgraphs as possible, a lower bound on the number of connecting edges can be found from

$$\min_{D} \sum_{i=1}^{q} \lambda_i (A+D) \ ,$$

the sum of the q largest eigenvalues of the incidence matrix with a diagonal matrix D of trace zero added.

5. Markov chains. Let an experiment have n possible outcomes $E_1,\ldots,E_n$ and perform a series of experiments. We have a Markov chain if the probability of a certain result E_k of an experiment depends only on what happend last time the same experiment was performed. p_{jk} is the probability of E_k occurring after E_j, and the series of experiments can be described by

$$\Pi_t = \Pi_{t-1} \cdot P$$

with Π_t the row vector of probabilities of outcome E_j at time t, and P the matrix of transition probabilities p_{jk}.

Clearly P is a nonnegative matrix with unit row sums. Its dominant eigenvalue is 1, and the corresponding (left) eigenvector describes the steady state of the chain. In fact, to obtain it corresponds to solving a homogeneous linear system of equations.

As opposed to the other examples we have considered, P is generally nonsymmetric. In realistic cases it is often very large and sparse. See the example given in Betteridge [3] concerning storage reallocation strategies in an operating system for a computer (fig. 1.3).

On some occasions also the smaller eigenvalues are of interest, mainly in order to judge the distance of a given state from the steady state. See e.g. W.J. Stewart [40].

Fig. 1.3 Markov chain transition matrix.
Different filling status in a storage allocation model (from Betteridge ref. 3).

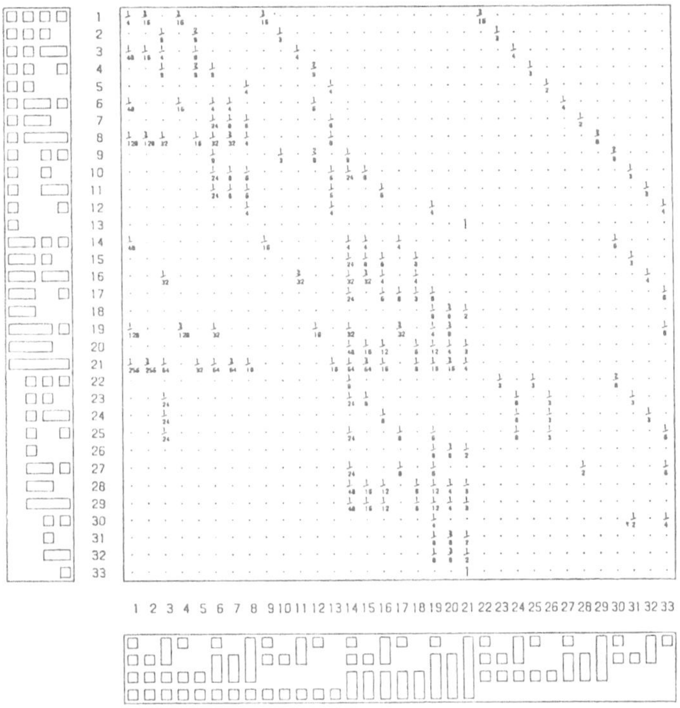

Fig. 9. Transition probability matrix ($N = 4$, best fit, non-relocating)

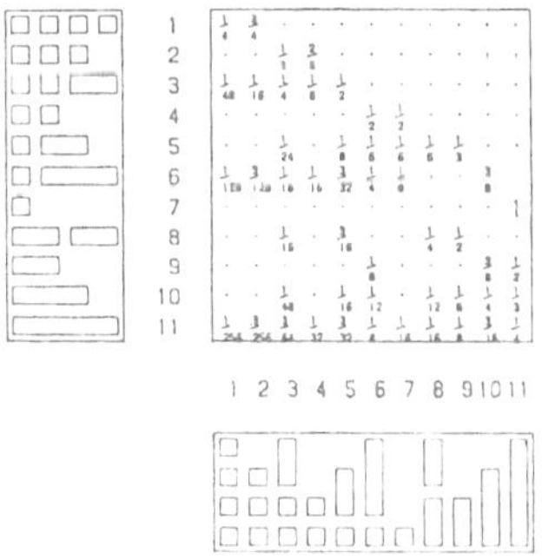

Fig. 10. Transition probability matrix ($V = 4$, relocating)

2. Transformation methods

In this section we will review the methods used for small dense eigenvalue problems in order to be able to determine which of these are also advantageous for large sparse problems. We will see that some of the decisive arguments for choice of method in the dense case no longer hold, which makes it necessary to start from the very beginning of the historical development.

Since finding an eigenvalue is the same as solving an n-th degree algebraic equation, all methods are necessarily iterative in nature. However, they differ in the amount of direct transformations performed on the matrix before an iterative process is begun. At one extreme is the direct iteration, or power method, which uses the matrix as it stands to find a sequence of vectors converging towards an eigenvector. On the other hand, it is also an easy matter to find the coefficients of the characteristic polynomial by means of a similarity transformation to companion form (See Wilkinson [43] pp.369, 405), and then apply a polynomial solver. It is a well-known fact that this is not a reliable technique and may give rise to eigenvalue approximations far from the true eigenvalues, even in cases where the matrix is well-conditioned with respect to the eigenproblem. We thus have to stop somewhere in between, and this is done in the following way:

1. We start with

$$(2.1) \qquad Ax = \lambda Bx.$$

Let us assume that A and B are symmetric and that B is positive definite and well-conditioned with respect to inversion. (We will make remarks on the nonsymmetric case now and then. There are methods to deal with singular B matrices; see e.g. Moler and Stewart [25]).

2. Factorize

$$(2.2) \qquad B = LL^T, \quad (1/6\ n^3)$$

L lower triangular, by means of symmetric Gaussian elimination or the Cholesky method. Form

$$(2.3) \qquad C = L^{-1}AL^{-T} \quad (2/3\ n^3)$$

to reduce (2.1) to

(2.4) $$Cy = \lambda y ,$$

a standard eigenvalue problem with the same eigenvalues and eigenvectors

(2.5) $$y = L^T x.$$

We have given the leading term in the arithmetic operation count in parentheses after each step. Since the number of additions and multi-- plications is roughly the same, we have listed that common number. Standard implementations of these algorithms are given in the Handbook of Linear Algebra, Wilkinson, Reinsch [19], and the discussion here is based on these implementations.

3. Transform

(2.6) $$C = PTP^T \quad (2/3\ n^3)$$

by means of an orthogonal similarity P into a tridiagonal matrix T. The transformation can be performed in a finite predetermined number of steps by means of the Householder method.

4. Perform an iterative algorithm to find the eigenvalues of T.

$$T_o = T$$

(2.7) $$T_{s+1} = Q_s^T T_s Q_s \quad (13n + n\ SQRT)$$

$$T_s \rightarrow D$$

where T_s are tridiagonal, Q_s orthogonal, and D diagonal with the eigenvalues as diagonal elements. The transformation matrices Q_s are almost always determined by means of the QR method.

Normally only a few iterations (3-5) are needed for each eigenvalue, so the total number of operations needed to compute all the n eigenvalues of a tridiagonal matrix is only a moderate multiple of n^2. It is therefore no significant saving to compute only one or a few eigenvalues instead of all n since (2.6) will in any event take an order of magnitude more time and has to be performed once in both cases.

5. a) If all eigenvectors are needed they are preferably computed by calculation of the combined transformation matrix

(2.8) $Y = PQ_1 \dots Q_s \quad (2/3n^3+4sn^2)$

followed by the calculation

$$X = L^{-T}Y \qquad (1/2n^3)$$

We see that these calculations, in spite of their simplicity, take a significant amount more time than the complete eigenvalue computation.

b) If one or a few eigenvectors are needed they are obtained by inverse iteration on T and then back-transforming with P and L^T. For each eigenvector we thus do:

$$T-\lambda_k I = LU \qquad (3n)$$

(2.9) $$z_k = U^{-1}e \qquad (3n)$$

(In some cases more iterations are needed here)

$$y_k = Pz_k \qquad (n^2)$$

$$x_k = L^{-T}y_k \qquad (n^2/2)$$

It might be appropriate at this point to list a few important properties of this algorithm. Together they are reason enough for it to outperform all other methods for the calculation of eigenvalues of dense stored matrices.

1. Orthogonal transformations are used throughout. This guarantees the numerical stability and insensitivity to rounding errors of the process.
2. The tridiagonal form is preserved while performing the transformations (2.7). This makes it necessary to do the time consuming (2.6) only once.
3. We have a shift strategy (not described here) that guarantees convergence after only a few iterations of (2.7) for each eigenvalue.
4. We have also a stable deflation that insures that every eigenvalue is computed once and only once.

There are other algorithms that have one or a few of these properties, but it is the fact that the Householder-QR algorithm has them all that is significant. So does the Jacobi algorithm using orthogonal transformations (1). The shift strategy is equivalent to Rayleigh quotient iteration (Parlett, Kahan [30]), a variant of Newton's method. However, it needs the solution of a linear system in each iteration, making the total number of operations a multiple of n^4.

We will now scrutinize the steps of the Householder-QR algorithm to see which modifications are needed to apply it to a sparse matrix where no n×n array can be stored for fast access.

If B is not just a diagonal matrix, we first have to consider the preliminary step 2. The factorization (2.2) is possible to perform whenever direct methods for the solution of a linear system with B as matrix are feasible. Then a sparse L can be obtained and stored, but the transformed matrix C is rarely sparse at all, not even when both A and B are tridiagonal from the beginning. When a sparse L is available it is on the other hand possible to perform matrix-vector multiplications with C in an efficient manner, since this amounts to a multiplication by A and one forward and one backward substitution by L.

Now let us assume that we have passed step 2 and have a large, sparse C. How likely is it that it shall be possible to find an efficient way of performing the reduction to tridiagonal form in step 3?

Let us look at a simple example of what happens when one step of Householder's method is performed.

Starting matrix:

$$A^{(0)} = \begin{bmatrix} x & x & & x & & x \\ x & & x & & x & \\ & x & x & x & & \\ x & & x & x & x & x \\ & x & & x & & \\ x & & & x & & x \end{bmatrix}$$

Pivot column.

$$A' = \left(I - \frac{UU^T}{K}\right)A^{(0)}, \quad K = \frac{1}{2}\,U^TU \ .$$

The sparsity pattern of U is that of the pivot column of $A^{(0)}$. If a_k and a_k' denote the kth column of $A^{(0)}$ and A', respectively, then

$$a_k' = a_k - \frac{UU^T}{K}\,a_k = a_k - \left(\frac{U^Ta_k}{K}\right)U, \quad 1 < k \leq 6$$

and we see from the expression in parentheses that fill-in takes place in the kth column if a_k and U have nonzero elements in at least one common position.

Moreover, the sparsity pattern of a_k' is the union of the sparsity patterns of a_k and the pivot column. Thus the sparsity pattern of A' is

$$\begin{bmatrix} x & x & & x & & x \\ x & & x & y & x & y \\ & x & x & x & & \\ x & & x & x & x & x \\ & x & & x & & \\ x & & y & x & y & x \end{bmatrix}$$

↑ No intersection

This is the same that would happen when solving an overdetermined linear system. Now let

$$A'' = A'\left(I - \frac{UU^T}{K}\right) .$$

Every row of A' that now intersects U is filled by the U elements and we find the sparsity pattern of A" is

$$\begin{bmatrix} x & x & & x & & x \\ x & z & x & y & x & y \\ & x & x & x & & z \\ x & z & x & x & x & x \\ & x & & x & & z \\ x & z & y & x & y & x \end{bmatrix}$$

Now we see that the next pivot column is full, filling it all up the next step.

We can pivot for sparsity only before starting the algorithm; any permutations applied afterwards will destroy the tridiagonal pattern already produced.

This simple study indicates that the chance of maintaining sparsity during reduction to tridiagonal form is much slighter than when performing Gaussian elimination for solving a linear system. There are other methods of similarity reduction, mainly Givens and Gaussian methods, which may be slightly better. A study by Duff and Reid [13], however, indicates that it is only on very rare occasions possible to save any substantial amount of storage space by performing sparse similarity reduction.

It is not even possible to retain band form in the matrix during the straightforward Householder method, but there is a modification due to Schwarz (See Handbook [19] p.273) which reduces the band width of a symmetric matrix by means of orthogonal rotations.

The remaining steps of the algorithm cause, oddly enough, no complications from a sparsity point of view. We could easily perform the QR algorithm (2.7) on a tridiagonal matrix that fills the memory, and eigenvector determination by means of inverse iteration (2.9) is also an easy matter.

For band matrices with a narrow band of filled elements, it is possible to apply the QR algorithm without first transforming the matrix into tridiagonal form. The time per iteration is roughly the same as that needed for solving a linear system by a band method, and the shifting and deflation strategy will also work. Because each iteration (2.7) is now more expensive, it is necessary to stop the process when the interesting eigenvalues have been found, often well before all n are computed. It is not possible to prescribe a certain order for the eigenvalues to appear, but a Sturm sequence test can be applied afterwards to determine whether more eigenvalues are to be found in a certain interval. See Handbook [19] p.266, for a standard implementation.

To sum up:
Matrices with a more general sparsity structure cannot be treated by similarity transformation methods of the type that are applied to dense stored matrices. We have to rely upon direct iterations where only use is made of the matrix-vector multiplication

$$y = Ax$$

and inverse iterations, where a sparse factorization makes it possible to solve linear systems

$$A-k_s I = LU$$

$$y = U^{-1}L^{-1}x.$$

3. Direct iterations: One-step methods

In this section we will discuss direct iterations which compute a sequence of vectors hopefully converging towards an eigenvector,

(3.1) $\quad x_1, x_2, \ldots, x_s \to u_k$ eigenvector,

and a sequence of eigenvalue approximations,

(3.2) $\quad \mu_1, \mu_2, \ldots, \mu_s \to \lambda_k$ eigenvalue.

We assume that the eigenvalues of (2.1)

$$Ax = \lambda Bx$$

are ordered so that

(3.3) $\quad \lambda_1 \leq \lambda_2 \leq \cdots \leq \lambda_n.$

Normally we seek λ_1 (or λ_n), and most methods give convergence towards an eigenvalue at the end of the spectrum, but we do not have any easy method of making sure that we really have found the end eigenvalue. Therefore the k in (3.1) and (3.2). For some problems, especially physical vibration problems, one can use the sign distribution of the eigenvector to determine whether the end eigenvalue has been found.

3.1 Properties of the Rayleigh quotient

For a given eigenvector approximation x, we determine the eigenvalue approximation μ as the Rayleigh quotient

(3.4) $\quad \mu(x) = x^T Ax / x^T Bx$

It is a well-known fact that all the eigenvalues are the only stationary values of $\mu(x)$, and a large class of eigenvalue algorithms can be regarded as methods for optimizing $\mu(x)$. In order to see how an optimization method works on the Rayleigh quotient, it is of interest to investigate some of its more important properties.

1. $\mu(x)$ is homogeneous in x,

(3.5) $\quad \mu(\alpha x) = \mu(x).$

Some kind of normalization can be applied if needed.

2. $\mu(x)$ is the number that makes μBx closest to Ax in the B^{-1}-norm for all choices of μ.

$$x^T\{(A-\mu B)x\} = 0$$

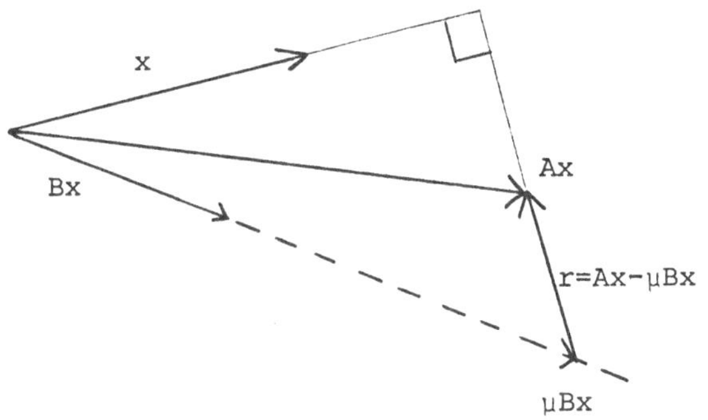

$B \neq I$

$B = I$

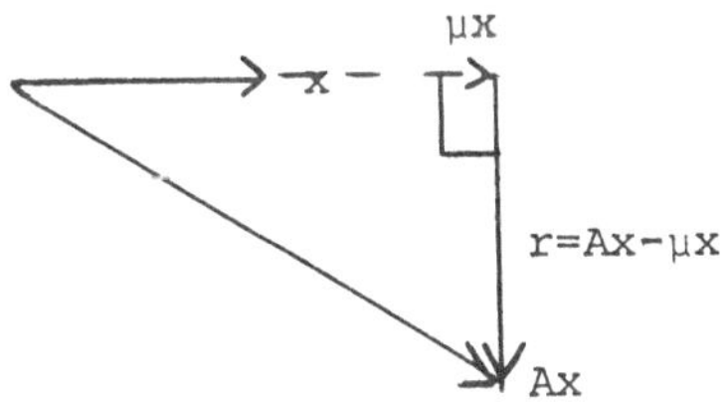

3. If x is close to an eigenvector u_k then $\mu(x)$ is still closer to λ_k:

Suppose x makes the angle θ to u_k,

$$x = \cos\theta\, u_k + \sin\theta\, v,$$

where all vectors are B-unit vectors,

$$u_k^T B u_k = v^T B v = x^T B x = 1,$$

and v is in the B-orthogonal complement of u_k

$$v^T B u_k = 0.$$

Then $\mu(x)$ is given by

$$\mu(x) = x^T A x = \cos^2\theta\, u_k^T A u_k + 2\cos\theta\sin\theta\, v^T A u_k + \sin^2\theta\, v^T A v$$

$$[A u_k = \lambda_k B u_k \Rightarrow v^T A u_k = 0]$$

$$\mu(x) = \lambda_k - \sin^2\theta(\lambda_k - v^T A v),$$

so the error in the eigenvalue approximation is proportional to the square of the error in the eigenvector approximation. This will be an important consideration in the analysis of several iterative methods.

When they are close to convergence, it can be assumed that the eigenvalue approximations are constant.

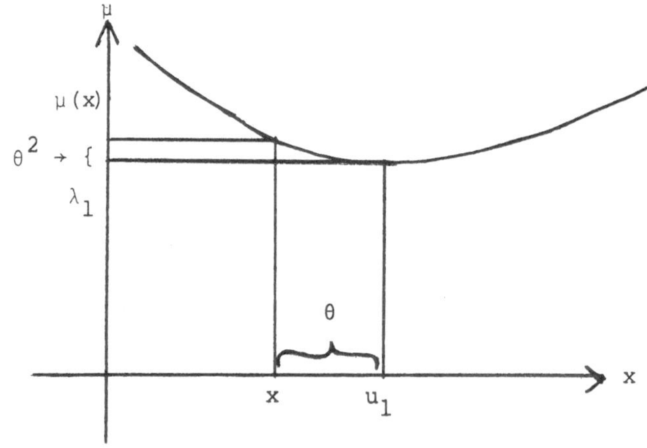

4. The gradient of $\mu(x)$ is a multiple of the residual:

$$(3.6) \qquad g(x) = \mu'(x) = \frac{2}{x^T Bx} (Ax - \mu(x)Bx)$$

Here we see that the stationary values of $\mu(x)$ are the eigenvalues.

5. The bigradient (or Hessian) of $\mu(x)$ is equal to the shifted matrix at the stationary points:

$$(3.7) \qquad H(x) = \mu''(x) = \frac{2}{x^T Bx} [A - \mu(x)B - g(x)(Bx)^T - Bx\, g(x)^T] \, .$$

We see that the homogeneity of $\mu(x)$ is reflected by the fact that

$$(3.8) \qquad x^T g(x) = 0, \quad H(x)x = -g(x).$$

6. The rate of convergence of most iterative optimization algorithms is dependent on the condition number of the Hessian. At the minimum point u_1 we have (3.7),

$$(3.9) \qquad H(u_1) = 2(A - \lambda_1 B)$$

and disregarding the singularity, which is appropriate to do, we get

$$(3.10) \qquad K_B(H) = \sup_{x \in S_1} \|Hx\|_{B^{-1}} / \inf_{x \in S_1} \|Hx\|_{B^{-1}} = (\lambda_n - \lambda_1)/(\lambda_2 - \lambda_1)$$

$$S_1 = \{x:\ x^T Bu_1 = 0,\ x^T Bx = 1\}$$

We see that $K_B(H)$ is dependent on the relative separation of λ_1 from the rest of the spectrum.

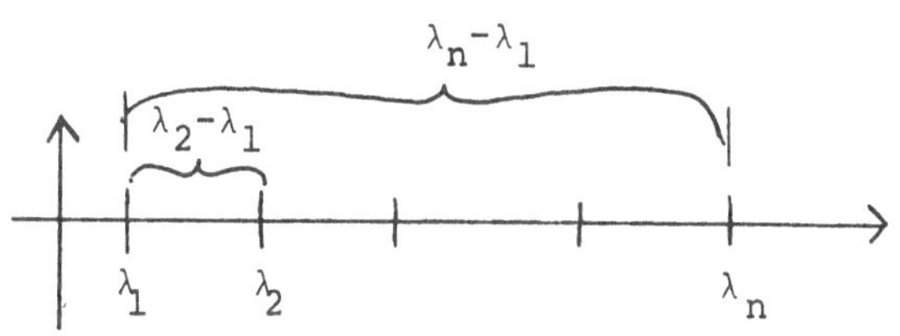

Level surfaces of $\mu(x)$ in the hyperplane orthogonal to u_1:

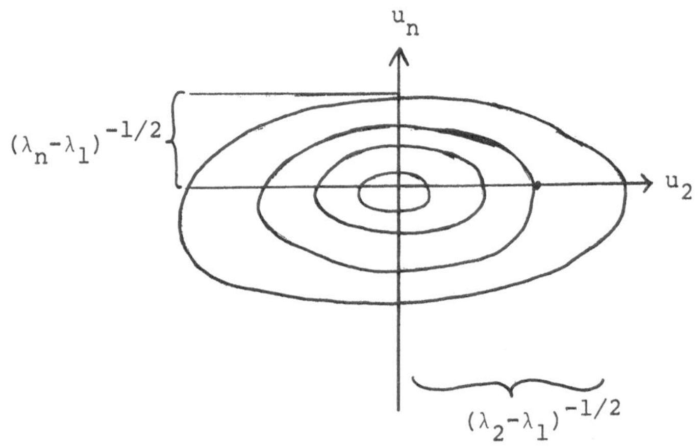

3.2 Simple iteration

We are now ready to state a few iterative methods. First we note that the simple iteration (power) method can be described in the context of minimizing the Rayleigh quotient.

The method of B-steepest descent applied to $\mu(x)$ is equivalent to the power method with shift chosen optimally.

B-steepest descent computes the next iteration x_{s+1} by

$$x_{s+1} = x_s - \alpha_s B^{-1} g(x_s)$$

$$= x_s - \alpha_s B^{-1}(A-\mu_s B)x_s \cdot 2/x_s^T B x_s$$

$$= -2\alpha_s/x_s^T B x_s \cdot B^{-1}(A-k_s B)x_s$$

$$k_s = \mu_s + (2\alpha_s)^{-1} x_s^T B x_s$$

which is the shifted power method. The shift k_s tends to a limit in the interval (λ_2, λ_n) when the algorithm converges towards u_1. See the classical textbook by Faddeev and Faddeeva [14] for further details.

When we now go further into the description of direct iterations for eigenvalue calculation, we find that there is a close analogy to the theory of iterative methods for linear systems Ax=b. There we study the quadratic form

$$(3.11)\qquad Q(x) = (Ax-b)^T A^{-1}(Ax-b)$$
$$= x^T Ax - x^T b - b^T x + b^T A^{-1} b$$
$$Q'(x) = 2(Ax-b)$$
$$Q''(x) = A$$

instead of the Rayleigh quotient.

For linear systems we can develop methods of two classes, stationary or norm-reducing methods

$$(3.12)\qquad x_{s+1} = Mx_s + k$$

with M chosen to give fast convergence, and nonstationary, or projection, methods which rely upon some kind of orthogonalization. The Gauss-Seidel and SOR methods are the most commonly known methods of the first class, while c-g (conjugate gradients) belongs to the second.

3.3 Norm-reducing methods

A norm-reducing method can be described by means of a splitting of the (shifted) matrix

$$(3.13)\qquad A-\mu_s B = V_s - H_s \qquad \mu_s = \mu(x_s)$$

$$(3.14)\qquad x_{s+1} = V_s^{-1} H_s x_s$$

In order to be applicable, the splitting should be chosen so that V_s is easy to invert while H_s should have as small norm as possible. In the linear systems case the rate of convergence is determined by the spectral radius,

$$\rho(V^{-1}H)$$

but now, since $A-\lambda_1 B$ is singular, this radius is always at least 1:

$$A-\lambda_1 B = V-H, \quad (A-\lambda_1 B)u_1 = 0$$
$$V^{-1}Hu_1 = u_1 - V^{-1}(V-H)u_1 = u_1 ,$$

so u_1, the eigenvector we seek, is an eigenvector of the iteration matrix with eigenvalue 1. We have to choose the splitting so that all the other eigenvalues of the iteration matrix are contained in the smallest possible circle inside the unit circle of the complex plane. Note that V and H need not be symmetric, so the eigenvalues of $V^{-1}H$ might be complex.

A natural choice for V_s is to take

(3.15) $$V_s = \omega^{-1}D_s - E_s$$

where

$$A-\mu_s B = D_s - E_s - E_s^T$$

divides the shifted matrix into diagonal lower and upper triangular parts. It corresponds practically and theoretically to the SOR method. Study fig. 7.4a and b at the end of the present report for a typical distribution of the eigenvalues of the iteration matrix for different choices of ω.

The overrelaxation factor is chosen in a way that is closely related to the linear systems case, and the method is applicable to the same class of matrices, mainly those emanating from finite difference or finite element approximations of partial differential equations. See Ruhe [34] for a more complete account.

It is easy to perform, working through the coordinates 1,...,n, and needs

$$6n+z$$

arithmetical operations for each iteration. We let z denote the total number of nonzero elements in $A-\mu_s B$.

If we remember the Gauss Seidel method for linear systems, we note that it works sequentially through the coordinates 1,...,n. For each coordinate p it finds the value of x_p that makes the p th equation satisfied. The same idea applied to the eigenvalue problem leads to the method of coordinate relaxation.

$$x_{s+1} = x_s + \xi_s e_p$$

(3.16) ξ_s chosen to minimize $\mu(x_{s+1})$.

It has been used for a long time because of its conceptual simplicity;

see e.g. Faddev and Faddeeva [14], or Shavitt et al. [37] for further discussion and references.

It is natural to consider overrelaxation also for this algorithm

(3.17) $$x_{s+1} = x_s + \omega \xi_s e_p .$$

This has been done by Schwarz [36].

We have proved that the SOR method is a first-order approximation of coordinate overrelaxation [34]. Their asymptotic behaviour is thus the same, but coordinate overrelaxation might give a slightly faster convergence in the beginning. However it needs somewhat more work in each iteration than SOR, namely

$$27n + z$$

arithmetical operations, so I believe the SOR method is to be preferred. Note that for difference or element approximation matrices, z is only about $5n$ or $10n$.

Before leaving these norm-reducing methods we mention that there are other splittings that may be used to advantage.

The SSOR splitting might also be applied. It will not give as fast convergence as SOR, but the fact that the iteration matrix now is a product of symmetric matrices makes it possible to let it be the basic iteration for an acceleration process as described by Axelsson [1] in the linear systems case. We do not yet have any experimental evidence on the virtue of this method. See also the monumental textbook of Young [45] on iterative methods for linear systems.

For a certain class of problems it is possible to make use of a natural splitting of the original mathematical problem into an invertible part and a perturbation of limited norm. In the Schrödinger equation example, considered in the introduction, we had

$$A = -\Delta_h + P$$

where $-\Delta_h$ is a finite difference analogue of the Laplacian and P is a diagonal matrix. We can then take

$$V_s = -\Delta_h + k_s I$$

and get V_s^{-1} by applying a fast direct Poisson solver (See e.g. Concus and Golub [9] for a closely related application). If P only varies moderately we can get an iteration that converges in few steps. See Ruhe [35] for a further account of this method.

4. Direct iterations: Nonstationary or semi-iterative methods

The direct iterations we have hitherto considered all only use the last vector computed in finding the next

(4.1) $x_{s+1} = \varphi(x_s)$.

They can be considered *one-step methods*. One can expect it to be possible to get a better result if the earlier history of the iteration process is taken into account,

(4.2) $x_{s+1} = \varphi(x_s, x_{s-1}, \ldots, x_1)$.

The term *semi-iterative* methods is often used in this case.

We will now see that it is in fact the case that semi-iteration will yield better approximations. First we will investigate how an optimal SI method will work, then how it can be realized by means of the Lanczos method, and last how simpler methods with equally good asymptotic properties are possible. Also now the theory is closely related to that for linear systems, but normally each linear system method has two counterparts for the eigenvalue problem: one (nonlinear) that uses the same concept in its development, and one (linear) which is simply application of the linear systems method to the homogeneous system

$$(A-\mu B)x = 0 .$$

We saw, for example, that the SOR method for linear systems corresponds to both coordinate overrelaxation (a nonlinear method) and the SOR method for $(A-\mu B)x=0$ (a linear method).

Problem type	Linear method	Nonlinear method
Linear system ($Ax = b$)	SOR	—
Eigenvalue ($(A-\mu B)x = 0$)	SOR	Coordinate overrelaxation

We also notice that the linear method generally has the same asymptotic behaviour as the nonlinear, while the nonlinear has a local behaviour that better corresponds to the original linear systems method.

4.1 Optimization over subspaces

To find an optimal semi-iterative method, suppose that we by any means have computed p vectors

(4.3) $$x_1,\dots,x_p$$

and also have

(4.4) $$Ax_1,\dots,Ax_p,\quad Bx_1,\dots,Bx_p$$

available. What will that tell us about λ_1 and u_1?

Since λ_1 is the global minimum of $\mu(x)$ it is reasonable to take the approximation

(4.5) $$\lambda^{(p)} = \min_{x \in \text{span}\,\{x_1,\dots,x_p\}} \mu(x)$$

and for $u_1^{(p)}$ the x that realizes that minimum. With (4.3-4) available, (4.5) can be readily computed as the solution of a pxp eigenvalue problem:

$$X = [x_1,\dots,x_p]$$

$$x = Xz, \qquad z \in R^p.$$

(4.6) $$\mu(x) = x^TAx/x^TBx$$

$$= z^TX^TAXz/z^TX^TBXz$$

$$= z^TEz/z^TFz$$

$$\lambda^{(p)} = \text{min eigenvalue of } E-\lambda F, \quad pxp.$$

If B=I then F=I provided that X is orthonormal.

It is worth noticing that the only thing that matters is the subspace

$$\text{span}\,\{x_1,\dots,x_p\},$$

not the individual vectors. This leads to several interesting consequences.

1. Any choice of basis is appropriate mathematically. Numerically it is of advantage if the vectors are chosen orthogonal since then (4.6) is as well-conditioned as the original problem and the subspace is well determined.

2. There is a sequence of subspaces, the *Krylov sequence*, which is natural to use as the basis of an iterative process. If B = I we define

(4.7) $A_s(x) = \text{span}\ \{x, Ax, \dots, A^{s-1}x\}$

$A_s(x)$ is obtained from $A_{s-1}(x)$ if we move out in the direction of the gradient of $\mu(x)$, as can be seen from (3.6). If nothing other than $A_{s-1}(x)$ and $A\ A_{s-1}(x)$ is known, this is the only choice we can make.

4.2 The Lanczos method

A semi-iterative method which is based on the optimization of $\mu(x)$ over a nested sequence of subspaces is a *Ritz method*, and we will now show that the Ritz method based on the Krylov sequence $A_s(x)$ is feasible to realize. It is done by means of the *Lanczos method* which successively constructs an orthogonal basis of the Krylov sequence of subspaces in such a way that the matrix A is represented by a *tridiagonal matrix* whose eigenvalues and eigenvectors are easily computed. (E in (4.6) is tridiagonal and F=I)

We choose to give a description of the Lanczos algorithm that closely follows the actual computations. This description pertains only to the case B=I. In order to apply the Lanczos algorithm to other cases, it is necessary that B can be factorized as in (2.2); that is, a linear system with B as matrix shall be solvable. See Weaver and Yoshida [42].

The starting vector x is given. We will have $A_s(x)$ spanned by

(4.8) $V_s = [v_1, \dots, v_s]$

if we take

$$v_1 = x/||x||_2$$

and compute for $k = 1, 2, \dots, s$

$$\alpha_k = v_k{}^T A v_k$$

(4.9) $$w_{k+1} = Av_k - \alpha_k v_k - \beta_{k-1} v_{k-1} \qquad (\beta_0 = 0)$$

$$\beta_k = ||w_{k+1}||_2$$

$$v_{k+1} = w_{k+1}/\beta_k \ .$$

Evidently

$$AV_s = V_sT_s + \beta_s v_{s+1}e_s^T$$

$$(4.10) \qquad T_s = \left[\begin{array}{ccc:c} \alpha_1 & \beta_1 & 0 & 0 \\ \beta_1 & \alpha_2 & \beta_2 & 0 \\ 0 & \beta_2 & \alpha_3 & 0 \\ \hdashline 0 & 0 & 0 & \alpha_s \end{array}\right]$$

We see that the vectors $v_1,\ldots v_s$ span $A_s(x)$. Thus $T_s = V_s^T A V_s$ represents the section of A in $A_s(x)$, and its eigenvalues will be the Ritz approximations to the eigenvalues of A.

It is worth noting that (4.9) implies that V_s is orthonormal, since assuming V_k is orthonormal we can prove that w_{k+1} is orthogonal to V_k. The choice of α_k implies $w_{k+1} \perp v_k$, and that of β_{k-1} that $w_{k+1} \perp v_{k-1}$ since

$$(4.11) \qquad v_{k-1}^T w_{k+1} = v_{k-1}^T Av_k - \beta_{k-1}v_{k-1}^T v_{k-1} = 0$$

$$\begin{aligned} \text{for } \beta_{k-1} &= v_{k-1}^T Av_k = v_k^T Av_{k-1} \\ &= v_k^T (w_k + \alpha_{k-1}v_{k-1} + \beta_{k-2}v_{k-2}) \\ &= v_k^T w_k \\ &= ||w_k||_2 \\ &= \beta_{k-1} \quad \text{chosen by (4.9)} \end{aligned}$$

and finally for $j < k-1$

$$\begin{aligned} (4.12) \qquad v_j^T w_{k+1} &= v_j^T Av_k = v_k^T Av_j \\ &= v_k^T (w_{j+1} + \alpha_j v_j + \beta_{j-1}v_{j-1}) = 0. \end{aligned}$$

This orthogonality implies that the algorithm stops after at most n steps.

When applying the Lanczos algorithm numerically, we cannot insure that the v vectors are orthogonal since the orthogonality depends on an inductive assumption for $j < k-1$ (4.12).

The algorithm was originally proposed as a means of tridiagonalizing a matrix (see Lanczos [24]), but this lack of orthogonality led to its being abandoned for the Householder method described is section 2. However, its use as an iterative method for $s < n$ was advocated by Paige [28] and has aroused a widespread interest in it. Paige makes an analysis of the effect of rounding errors, indicating in which cases loss of orthogonality occurs. Kahan and Parlett [23] show elegantly that it is possible to stop long before any β_k is negligible and still get very accurate eigenvalue and eigenvector approximations.

Let us see how this will happen. From (4.10) we see that

(4.13) $$AV_s - V_sT_s = \beta_s v_{s+1} e_s^T = R_s.$$

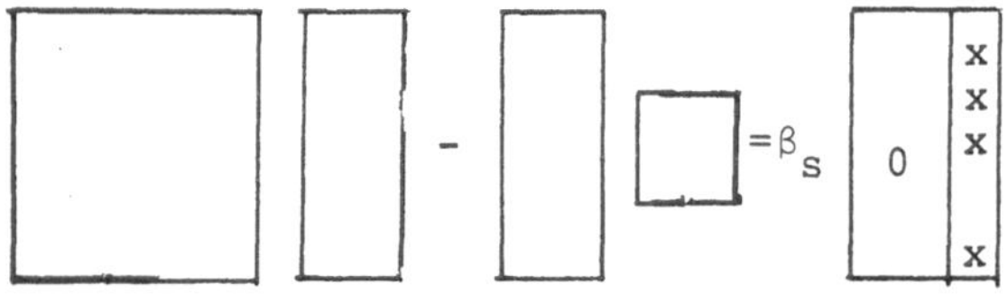

Kahan has proved that there is a correspondence between the eigenvalues of T_s and some of those of A such that

(4.14) $$|\lambda_i(T_s) - \lambda_i(A)| \leq ||R_s||.$$

Now look at the special appearance of R_s. Evidently $||R_s||$ is small whenever it happens that β_s is small, and this is an evident stopping criterion for the algorithm. However, in practice it seldom happens that β_s ever gets small, and then it is interesting to note that <u>some</u> of the eigenvalues often are much more accurately approximated than indicated by the size of β_s. This is because R_s has all its elements in its last column.

Let z be an eigenvector of T_s:

$$T_s z - \lambda z = 0.$$

Then transforming back to the basis of A we get

$$x = Vz ,$$

$$(4.15)\qquad \begin{aligned} Ax - \lambda x &= AVz - \lambda Vz \\ &= AVz - VT_s z \\ &= \beta_s v_{s+1} e_s^T z , \end{aligned}$$

$$||Ax-\lambda x|| = |\beta_s||(z)_s| ,$$

and we have multiplied β_s by the last element of z. Consequently, all eigenvalues of T_s whose vectors have small last elements have smaller errors than expected from β_s. It is an interesting fact that tridiagonal matrices arising from practical problems usually have the property that the extreme eigenvalues have eigenvectors with very small last elements.

When the computations are performed with roundoff, these results are more difficult to prove. First, all elements of R_s are not concentrated in the last column since all columns contain roundoff. Second, V is no longer exactly orthogonal; we have to provide expressions for what happens in mildly nonorthogonal cases and criteria to stop when nearly linearly dependent vectors occur. See the above mentioned work by Kahan and Parlett [23] or the dissertation by Paige [27].

It is also possible to give an a priori bound on the accuracy of the approximation $\lambda^{(s)}$ obtained after s steps with the Ritz method. It takes only the accuracy of the starting vector and the condition (3.10) of the eigenvalue problem into account. We can prove that

$$(4.16)\qquad \lambda^{(s)} - \lambda_1 \le \left(\frac{\tan\theta}{T_s(\sigma)}\right)^2 (\lambda_n-\lambda_1)$$

where θ is the angle between the starting vector and u_1, and T_s is the Čebyšev polynomial

$$T_s(\xi) = \cosh\,(s \cosh^{-1}\xi)$$

evaluated at

$$\sigma = 1 + 2\,\frac{\lambda_1-\lambda_2}{\lambda_2-\lambda_n} = \frac{K+1}{K-1} \ge 1 + 2K^{-1}$$

with K given by (3.10).

For large s we can estimate

$$(T_s(\sigma))^{1/s} \simeq \frac{\sqrt{K}+1}{\sqrt{K}-1} \quad .$$

We see from (4.16) that $\lambda^{(s)}$ converges towards λ_1 at a rate determined by the reciprocal of this value. For large K we get, taking the logarithm,

$$R \simeq 4K^{-1/2}.$$

This is a considerable improvement over steepest descent where

$$R \simeq 4K^{-1},$$

if we remember that K is very large for problems of a practical scale.

For 2-cyclic matrices (property A) we can prove a similar result for SOR, but note that this result for Lanczos bounds the worst possible case. Normally convergence occurs at a much faster rate.

Let us see how much work is needed to perform s steps of the Lanczos method. The matrix V_s will be dense, but we note that only the two last vectors are needed during the computation (4.9). These vectors can then be written on some sequential storage medium and then taken back when an approximate eigenvector of T_s is transformed back to the basis of A. The iteration (4.9) will need

$$(z + 5n)s$$

arithmetic operations. We then find one eigenvalue and one eigenvector of T_s by means of bisection and inverse iteration in

$$2st + 6s$$

operations (t number of bits accuracy) and then transform back into the basis of A in

$$s \cdot n$$

operations. Compare this to the norm-reducing methods considered in section 3, and note that we only need slightly more operations than the simplest SOR method (3.15) and fewer than coordinate relaxation (3.16), provided that $s \ll n$ and that the write-read of v-vectors does not cause any prohibitive overhead.

4.3 c-g optimization of the Rayleigh quotient

We have now developed the Lanczos algorithm and seen that it is the Ritz method applied to the Krylov sequence of subspaces. The corresponding method for a linear system of equations is the conjugate gradient (c-g) method. It is most often described by means of the iterates x_s, search directions p_s, and residuals r_s the following way:

$$x_{s+1} = x_s + a_s p_s$$

$$p_s = -r_s + b_{s-1} p_{s-1}$$

$$r_s = Ax_s - b,$$

where the parameters a_s and b_s are determined by

$$\text{(4.17)} \quad \begin{aligned} a_s &= -p_s^T r_s / p_s^T A p_s \\ b_s &= r_{s+1}^T r_{s+1} / r_s^T r_s . \end{aligned}$$

a_s insures that $Q(x) = r^T A^{-1} r$ (3.11) is minimized in the direction p_s, and b_s that the search directions p_s are A-conjugate.

It is interesting to note that this algorithm is equivalent to the Lanczos algorithm; the r_s vectors in c-g have the same directions as the v_s in Lanczos (4.9), provided that c-g is applied to the system

$$Ax = x_0 .$$

The Lanczos vectors v_s are independent of a shift of A (the only thing that happens is that T_s is shifted correspondingly), so we could as well regard a shifted matrix also in c-g:

$$(A - kI)x = x_0 .$$

We say this so that the relation to the c-g minimization algorithm applied to $\mu(x)$ shall be clearer.

The c-g algorithm has been in use for unconstrained minimization, also of functions other than quadratic forms. It is the simplest of the so-called quasi-Newton methods. It has the advantage that no nxn matrix need be stored and is thus the only one applicable to large sparse eigenvalue problems. Application of c-g to $\mu(x)$ was apparently first proposed by Bradbury and Fletcher [6] and applied extensively by Fried [15] [16] and Geradin [17], in both cases in a finite element context.

In contrast to Lanczos, c-g can as easily be applied to the general $Ax = \lambda Bx$ problem, so we describe it in that context. With a suitable starting x_0 we compute (compare (4.16)).

$$\text{(4.18)}\qquad \begin{aligned} x_{s+1} &= x_s + a_s p_s \\ p_s &= -g_s + b_{s-1} p_{s-1} \end{aligned}$$

with the gradient g_s given by (3.6),

$$g_s = g(x_s) = \frac{2}{x_s^T B x_s} (Ax_s - \mu_s B x_s),$$

and b_s chosen precisely as (4.17),

$$\text{(4.19)}\qquad b_s = g_{s+1}^T g_{s+1} / g_s^T g_s .$$

The step length parameter a_s, on the other hand, is usually chosen so that $\mu(x)$ is minimized in the direction of p_s. This amounts to solving a second-degree equation which in fact is the characteristic equation of $A-\lambda B$ restricted to the subspace spanned by x_s and p_s. We have

$$\mu(x+\xi p) = \frac{x^T Ax + 2\xi p^T Ax + \xi^2 p^T Ap}{x^T Bx + 2\xi p^T Bx + \xi^2 p^T Bp}$$

and the condition for a minimum is

$$a\xi^2 + \xi - \eta = 0$$

$$\text{(4.20)}\qquad \begin{aligned} \eta &= -p^T(A-\mu B)x / p^T(A-\mu B)p, \quad \mu = \mu(x) \\ a &= p^T B(x+\eta p) / x^T Bx \end{aligned}$$

We note the close correspondence between η here and a_s in (4.17). It is the same expression with r_s and A replaced by $g(x)$ and $A-\mu B$, respectively. Further, the first-order approximation to ξ is η, so asymptotically we get the same step lengths as when solving a linear system with $A-\mu B$ as matrix.

We will thus have reason to expect that the asymptotic behaviour of the c-g minimization algorithm (4.18-20) is the same as Lanczos. It is considerably more simple to implement since only three n-vectors need be stored. The number of arithmetic operations for one iteration is

$$z + 12n$$

for the $A-\lambda B$ problem, comparable to Lanczos, and in between SOR and coordinate overrelaxation.

Note that this c-g minimization algorithm does not need any factorization of B (2.2) as opposed to the Lanczos algorithm. It can thus be applied to a wider range of cases.

To sum up the developments in this section, we see that the picture of generalizations of projection methods to the eigenvalue problem will be:

Problem type	Linear method	Nonlinear method
Linear system ($Ax = b$)	c-g	—
Eigenvalue ($(A-\mu B)x = 0$)	c-g optimization of $\mu(x)$	Lanczos algorithm ($B = I$)

Compare to the corresponding picture for norm-reducing (one-step) methods given in the beginning of this section.

5. Direct iterations: Finding intermediate eigenvalues

We now turn to the problem of finding more than one eigenvalue by direct iteration. We have the following three options:

1. Deflation methods
2. Simultaneous methods
3. Spectral transformation methods

5.1 Deflation methods

A deflation method finds the next eigenvalue by orthogonalising all iterates to the eigenvector already computed.

Instead of $y = Ax$ it computes

(5.1) $$y = Ax - ((Ax)^T u)u = PAx,$$

where u is the computed eigenvector and

$$P = I - uu^T. \quad \text{(projection matrix)}$$

If p vectors already are computed, we use modified Gram Schmidt to find y:

$$w_1 = Ax$$

(5.2) $$w_{k+1} = w_k - (w_k^T u_k) u_k \qquad k = 1,\dots,p$$

$$y = w_{p+1}\,,$$

and we see that 2 pn extra arithmetical operations are needed for each iteration. The operation counts given earlier will hold with z replaced by z + 2pn, and we see that no large numbers p are allowed if the algorithm is to be effective.
The new matrix

(5.3) $$PA = PAP$$

is effectively the matrix A restricted to the subspace orthogonal to span $\{u_1,\dots,u_p\}$, and the rates of convergence for the iterative algorithms are determined by

$$K_p = \frac{\lambda_n - \lambda_{p+1}}{\lambda_{p+2} - \lambda_{p+1}} .$$

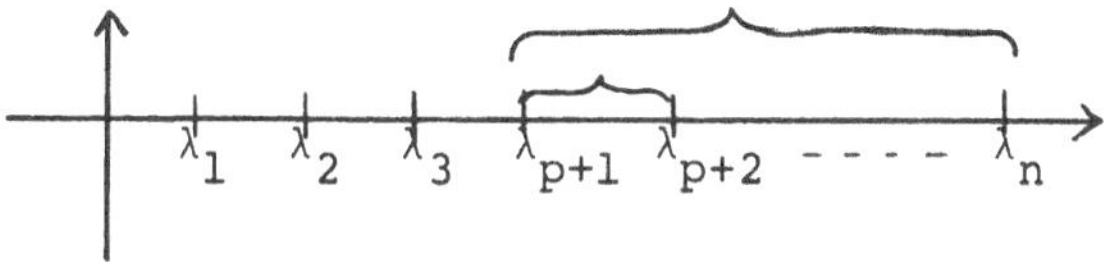

Compare to (3.10) and note that clustering eigenvalues will effectively ruin the convergence.

5.2 Simultaneous methods

This has been the reason for abandoning true deflation methods and instead using simultaneous methods which iterate on a few, say m, vectors simultaneously in order to find an m-dimensional subspace where μ is optimized. Properly implemented, such a method gives convergence to $\lambda_1, \ldots, \lambda_m$ with the rates determined by

(5.4) $\quad K_{r,m} = (\lambda_n - \lambda_r)/(\lambda_{m+1} - \lambda_r) \quad$ for λ_r.

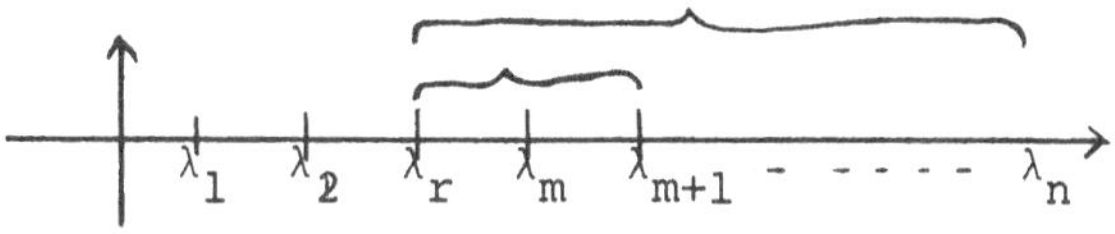

If we are interested in the p first eigenvalues we might choose $m > p$ and get good convergence even if the p first eigenvalues are clustered.

Simultaneous iteration has been developed into a reliable standard program by Rutishauser in Handbook [19]. Besides iterating on m vectors simultaneously it involves

1. Orthogonalization of the vectors at intervals when they are likely to become linearly dependent.
2. Finding eigenvalue approximations in the m-dimensional subspace as in (4.5-6) above.
3. Avoiding iterating on vectors that have already converged, effectively including a deflation strategy.
4. Acceleration of convergence by means of Čebyšev polynomials.

This program is still the standard iterative algorithm against which all alternative strategies are to be tested.

Simultaneous Lanczos has been proposed by several people (see Cullum and Donath [10], Underwood [41] and Kahan and Parlett [22]), normally termed Block-Lanczos. It can be implemented in several ways; the best and simplest is apparently the following, yet unpublished, variant:

1. Start with $v_1, \ldots, v_m$ orthonormal vectors.
2. Now we are going to get

   $AV = VT + E,$

   T 2m+1 band matrix.

   One column in this:

   $$Av_s = \sum_{k=1}^{m} v_{s-k} t_{s-k,s} + \sum_{k=0}^{m} v_{s+k}, t_{s+k,s}$$

   As in (4.9) we formulate it:

$$(5.5)\quad \begin{cases} w_{s+m} = - \sum_{k-1}^{m} v_{s-k} t_{s-k,s} + Av_s - \sum_{k=0}^{m-1} v_{s+k} t_{s,s+k} \\ t_{s,s+m} = \|w_{s+m}\|_2 \\ v_{s+m} = w_{s+m} / t_{s,s+m} \end{cases}$$

We get the elements in T successively from the top. In the first sum

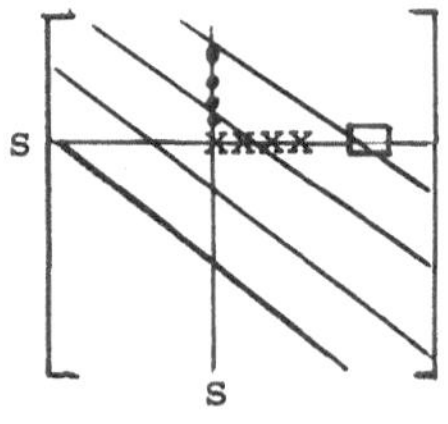

the already computed elements above t_{ss} are used. After that, the elements in the s-th row are computed; their values are given by the condition that w_{s+m} shall be orthogonal to $v_s, v_{s+1}, \ldots, v_{s+m-1}$. Last, the element $t_{s,s+m}$ is obtained as the norm of w_{s+m}. If that norm is small, Av_s is no longer linearly independent of $v_s, \ldots, v_{s+m-1}$, and we have to decrease the band width by one. From a numerical point of view, it is important in which order the computations are performed. A modified Gram Schmidt approach where the vector w_{s+m} is successively orthogonalized to each new v_{s+k} is much to be preferred.

We see that 2m+1 vectors are needed in store during the process.

3. The eigenvalues of T can now be computed by band QR or inverse iteration.

Each step of (5.5) takes

$$z + 3(m+1)n$$

arithmetic operations, after which the time for band QR is proportional to

$$sm^2.$$

The increase in computer time requirements, compared to the simple Lanczos method, is comparable to the extra cost involved when deflating against already computed vectors (5.2). The better convergence properties of the simultaneous method make it to be preferred.

In fact, the simultaneous method has to be combined with deflation. If there is an eigenvector in the span of $v_1,\dots,v_m$, then w_{2m} will necessarily be zero, causing a band width reduction in (5.5).

Underwood [41] has proved that this block-Lanczos method has the same rate of convergence as ordinary Lanczos (4.16) with the gap between λ_1 and λ_2 replaced by the gap between λ_r and λ_{m+1}, so the relevant condition number is (5.4) and we get the bound

$$\lambda_r^{(s)} - \lambda_r \le \left(\frac{\tan\theta}{T_s(\sigma_r)}\right)^2 (\lambda_n - \lambda_r)$$

$$\sigma_r = \frac{K_{r,m}+1}{K_{r,m}-1}$$

θ smallest angle between span $\{v_1,\dots,v_m\}$ and span $\{u_1,\dots$ $\dots,u_m\}$

s number of block iterations.

5.3 Spectral transformations

Both deflation and simultaneous methods find the eigenvalues starting at the end of the spectrum moving inwards. Most often it is also these eigenvalues that are of interest. For instance, in a finite difference approximation to a physical system the lowest eigenvalues correspond to the physically relevant modes of oscillation, while the higher are more contaminated by truncation errors.

If some eigenvalues in the middle of the spectrum are needed, we are in much worse a situation. In the case B=I, we can work with

$$(5.6) \qquad C = (A-kI)^2.$$

C has identical eigenvectors to A, but the eigenvalues closest to k of A correspond to the smallest eigenvalues of C .

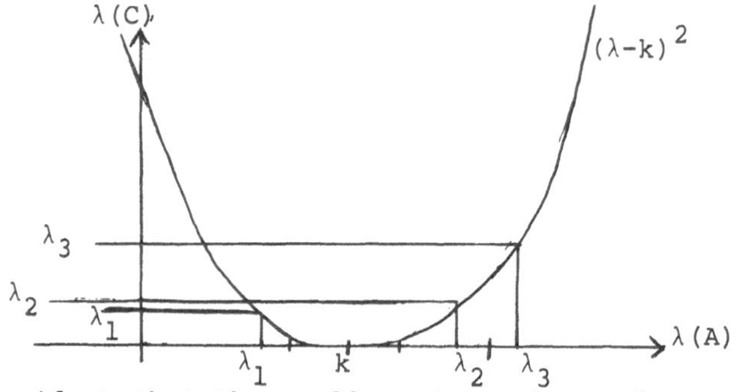

It is evident that the smaller eigenvalues of C are very badly separated compared to those of A. Suppose that the shift k hits λ_j exactly. Then

$$(5.7) \qquad \begin{aligned} \lambda_1(C) &= 0 \\ \lambda_2(C) &= \min_i (\lambda_i-\lambda_j)^2 \\ \lambda_n(C) &= \max_i (\lambda_i-\lambda_j)^2, \end{aligned}$$

virtually squaring the condition number K (3.10) which governs the rate of convergence of the iterative processes we have discussed. If we need 30 iterations for an end eigenvalue, we will need 1000 to get one in the middle by means of C.

Methods have been proposed (See e.g.Rodrigue [32]) which, instead of μ (3.4), try to minimize

$$(5.8) \qquad r(x) = ||Ax-\mu(x)Bx|| \,/\, ||x||$$

which is nonnegative and zero at the eigenvectors. It is easy to see that if such a method converges towards λ_j then the rate is the same as that of the shifted method (5.6) with $k=\lambda_j$. The same remarks on the slow rate, consequently, will apply also to these methods.

5.4 Lanczos' method

The Lanczos method is usually applied only to find a few eigenvalues in the end of the spectrum. However, at least mathematically, it may well happen that good approximations will be obtained for eigenvalues in the middle provided that other eigenvalues of T_s are chosen as approximations. The a posteriori error bound (4.15) is as well applicable to middle eigenvalues of T_s while the a priori bound (4.16) has to be replaced by something else. The Čebyšev polynomial in the denominator has to be replaced by the value in λ_j of a polynomial with a prescribed bound on its value in the rest of the spectrum:

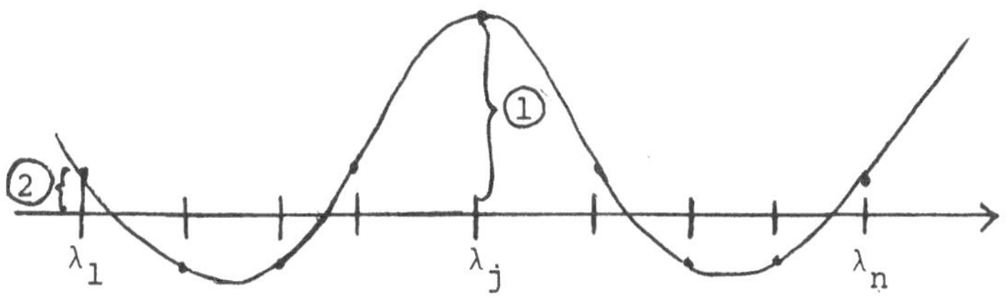

①/② should be maximized over s-degree polynomials.

A proof of this fact has been announced by Cline, Golub & Platzman [8].

Clearly it will be impossible to obtain a quotient as large in the middle as in the end, and poor separation of λ_j from the rest of the spectrum will worsen the results even more.

Very scant practical experience with this type of algorithm has been reported, but it is my belief that the convergence will slow down, possibly as much as when the spectral transformation algorithms for middle eigenvalues described earlier are applied.

6. Inverse iterations

As mentioned in the section on dense matrices, the methods used there were based on similarity transformations of the matrices. We found that it was very unlikely that sparsity of a matrix could be retained to any extent during such transformations and directed the study towards iterative methods.

We are now going to study the intermediate case; that is, matrices where we can find an inverse, or more properly, solve a linear system conveniently, but not perform similarity transformations. Inverse iteration techniques are then possible and most often preferable. We will see that the Newton method applied to a nonlinear equation formulation of the eigenvalue problem is realized by inverse iteration with changing shifts, Rayleigh quotient iteration (RQI). RQI has the same local convergence as QR and needs much fewer iterations than direct iterative methods.
The Newton method applied to the nonlinear system of equations

$$f(y) = 0$$

computes the sequence of approximations

$$y_{s+1}=y_s-F(y_s)f(y_s) \tag{6.1}$$

where $F(y_s)$ is the matrix of first derivatives of f.
We can reformulate the eigenvalue problem (2.1) as the following n+1 dimensional system:

$$\begin{aligned} f_1 &\equiv (A-\lambda B)x = 0 \qquad & y_1 &= x \\ f_2 &\equiv x^TBx-1 = 0 \qquad & y_2 &= \lambda \end{aligned} \tag{6.2}$$

and obtain the following (n+1) x (n+1) dimensional linear system in each iteration of (6.1),

$$\begin{bmatrix} A-\lambda_sB & -Bx_s \\ 2x_s^TB & 0 \end{bmatrix} \begin{bmatrix} x_{s+1} - x_s \\ \lambda_{s+1} - \lambda_s \end{bmatrix} = -\begin{bmatrix} (A-\lambda_sB)x_s \\ x_s^TBx_s-1 \end{bmatrix}$$

$$\begin{cases} (A-\lambda_sB)x_{s+1} = (\lambda_{s+1}-\lambda_s)Bx_s \\ 2x_s^TBx_{s+1} - x_s^TBx_s - 1 = 0 \end{cases} \tag{6.3}$$

Let us assume that the starting vector is B normalized so that the second equation of (6.2) is satisfied at the outset. Then we multiply the first equation of (6.3) by x_{s+1}^T and see that

$$\lambda_{s+1} = \mu(x_{s+1})$$

the Rayleigh quotient (3.4) at x_{s+1}.

Most often we renormalize x_s in every step and then we get

$$\begin{cases} (A-\mu_sB)z_{s+1} = Bx_s \\ x_{s+1} = z_{s+1}/||z_{s+1}||_B \\ \mu_{s+1} = \mu(x_{s+1}) \end{cases} \tag{6.4}$$

which is inverse iterations with Rayleigh quotient shifts.

To perform (6.4) we need to solve a linear system with a new shift in each iteration. In many cases this will be easy to do; the sparsity pattern is the same for all the systems.

The convergence of RQI was studied by Ostrowski in a series of papers [27]; see also Parlett and Kahan [30]. Convergence will almost always occur, and the rate will be cubic in the symmetric case.

It is worth noting that in the symmetric case we have a global convergence result for RQI, [30],[44].

Suppose x_s is B-normalized and consider

$$r_s = (A-\mu_s B)\, x_s.$$

Then

$$\begin{aligned} r_{s+1} &= (A-\mu_{s+1}B)\, x_{s+1} \\ &= (A-\mu_s B)\, x_{s+1} + (\mu_s-\mu_{s+1})\, Bx_{s+1} \\ &= c_s\, Bx_s + (\mu_s-\mu_{s+1})\, Bx_{s+1}. \end{aligned}$$

Now the normalization constant c_s is, by (6.4),

$$c_s = x_s^T (A-\mu_s B)\, x_{s+1} = r_s^T x_{s+1} = ||r_s||_{B^{-1}} \cos\theta,$$

where θ is the angle between $B^{-1/2} r_s$ and $B^{1/2} x_{s+1}$.

Furthermore, premultiplying (6.4) by x_{s+1}^T:

$$c_s\, x_{s+1}^T B x_s = x_{s+1}^T (A-\mu_s B)\, x_{s+1} = \mu_{s+1}-\mu_s,$$

so

$$||r_{s+1}||^2_{B^{-1}} = ||r_s||^2_{B^{-1}} \cos^2\theta - (\mu_{s+1}-\mu_s)^2 \le ||r_s||^2_{B^{-1}}.$$

Consequently the norms of the residuals decrease, and it can be proved that they almost always tend to zero. The convergence of μ_s is furthermore cubic.

Note that the matrix of (6.4) is not positive definite since μ_s is always inside the range of eigenvalues of $A-\lambda B$. If solution is performed by means of Gaussian elimination, one might record the number of negative pivot elements. It will give the number of eigenvalues smaller than μ_s and insure that we do not miss any eigenvalue of interest.

It might also be noted that the unnormalized solution z_{s+1} of (6.4) will have a very large norm when μ_s is close to convergence. As the analysis of Wilkinson shows, this is of no harm. The angle between x_{s+1} and the invariant subspace of λ_k is what matters, and that angle is small precisely when z_{s+1} will have a very large norm. See Wilkinson [43] and [44].

It is now and then proposed that $A-\mu_s B$ should be deflated so that we get an exactly singular system to solve. See e.g. Cline et al. [8] where

(6.5) $$C = (I-Bx_s x_s^T)\ (A-\mu_s B)\ (I-Bx_s x_s^T)$$

replaces $(A-\mu_s B)$ in (6.4). This corresponds to solving (6.2) for δ_s as it stands. Please note that the matrix $H(x)$ of the system in (6.2) is not sparse; its last two terms are generally full, though of rank 1.

It has been proposed that an iterative method suitable for indefinite symmetric matrices should be used to solve the linear system (6.4). See [8], where only the case $B=I$ is considered. The natural choice of method is a variant of the Lanczos method proposed by Paige and Saunders, [29], which computes the tridiagonal matrix T_s as (4.10) and successively updates a solution of the tridiagonal system. However, the solution is then found as a linear combination of the Krylov sequence started at x_s and could not be better as an eigenvector than the eigenvector corresponding to an appropriately chosen intermediate eigenvalue of T_s. I suspect that no inverse iteration based on iterative methods for the solution of linear systems is better than the direct iterative methods for the eigenproblem we have discussed in sections 3-5. This suspicion is founded on the fact that all iterative methods find their solution as a linear combination of Krylov vectors,

$$x,\ Kx,\ K^2x, \dots\ K^{s-1}x\ ,$$

of some matrix K closely related to $A-\mu B$. Especially we have no reason to expect that we will be able to derive good methods for intermediate eigenvalues which do not require that we really solve a linear system with $A-\mu_s B$ as its matrix.

This might be an overly pessimistic view of the state of affairs. If the intermediate eigenvalues are separated so that

$$\left(\frac{\lambda_n-\lambda_1}{\lambda_{k+1}-\lambda_k}\right)^2 \sim \frac{\lambda_n-\lambda_1}{\lambda_2-\lambda_1}$$

they are as easy to compute as λ_1 by the spectral transformation methods outlined in section 5 ((5.6) or (5.8)).

In some situations it might be considered too costly to refactorize the matrix $(A-\mu_s B)$ in (6.4) for a new μ_s in every iteration, but use a fixed factorization of

(6.6) $\quad A - kB = LU$

for several iterations. If just one value of k is chosen this can be considered as a spectral transformation replacing

$$\lambda_i \text{ by } \rho_i = 1/(\lambda_i - k),$$

making the ρ_i corresponding to eigenvalues close to k large in magnitude. We can then apply any direct iteration method to find the extreme eigenvalues of this well separated problem in order to get good estimates to the eigenvalues of the original problem that are close to the shift k. This technique can be combined with simultaneous iteration and periodic updating of the shift; see e. g. Jensen [21] or Underwood [41].

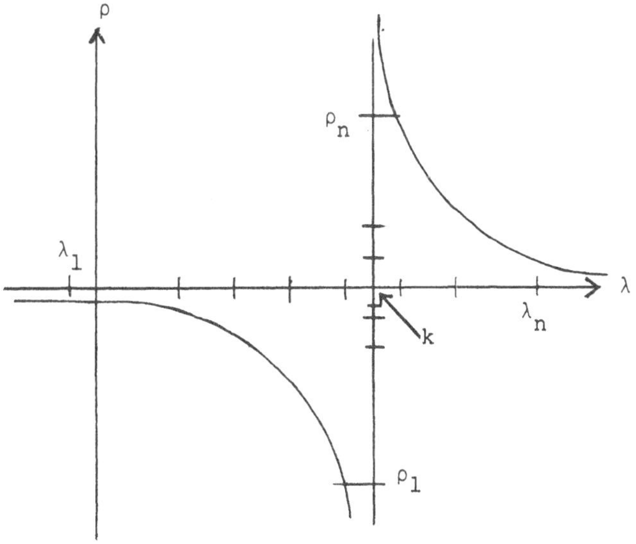

I have as yet quite limited experience with the application of inverse iterative methods to large sparse eigenvalue problems and have not been able to find any relevant comparisons between direct and inverse iterations in the literature. The only general advice that can be given is that if an explicit inversion procedure for the matrix at hand is available it should be used, since the inverse iterations have so much better convergence properties than even the most sophisticated direct iterations that it justifies the extra cost in each iteration. We have yet to wait for any quantitative support for this advice.

7. Numerical examples

In this section we will include three small, yet typical, numerical examples that give us a flavor of how the methods we have described work in practice. The first is a finite difference matrix arising from an elliptic equation, the second is a sparse symmetric matrix of a more irregular shape, while the third is a band matrix with a very poor separation between the smallest eigenvalues. In all three cases B=I. We have as yet rather scant experience with ill-conditioned B-matrices, and the well-conditioned ones behave the same as I.

The first example is the 5-point finite difference approximation to the two-dimensional Laplace equation, which yields a block tridiagonal matrix:

$$A = \left.\begin{bmatrix} B & -I & 0 & 0 \\ -I & B & -I & \\ & & & \\ 0 & & & B \end{bmatrix}\right\} nb$$

$$B = \left.\begin{bmatrix} 4 & -1 & & 0 \\ -1 & \ddots & \ddots & \\ & \ddots & \ddots & -1 \\ 0 & & -1 & 4 \end{bmatrix}\right\} b$$

Its eigenvalues are

$$\lambda_{ij} = 4\left(\sin^2 \frac{\pi i}{2(nb+1)} + \sin^2 \frac{\pi j}{2(b+1)}\right) \qquad \begin{matrix} i = 1,2,\ldots,nb \\ j = 1,2,\ldots,b \end{matrix}$$

and if we take n = 300 (nb = 15, b = 20) we get the condition number K = 120 (3.10) in either end of the spectrum. Studying fig.7.1 which gives a diagram of the convergence for this matrix, we see that the algorithms fall into three categories. Steepest descent is slowest, while SOR, Lanczos and c-g give approximately the same rate of convergence. For this well-conditioned problem the advantage of using the Lanczos algorithm is not very pronounced; the c-g algorithm gets full accuracy in the eigenvalue after about 75 iterations. Far fewer than n iterations were needed since the matrix is a direct sum of two commuting lower order matrices. In this case the simultaneous iteration [19] needed 113 iterations on 5 vectors to get the 2 lowest eigenvectors.

In order to show the behaviour of the algorithms on matrices that are not of difference type, we list results on a matrix we have obtained by taking the upper triangle of the 54x54 CURTIS matrix, setting all filled nondiagonal elements = -1 and the diagonal elements as the absolute value of the sum of the nondiagonal ones, except a_{11} and a_{nn} which are set as double that value. The eigenvalues are clustered in the lower end of the spectrum. We have

$$\lambda_1 = 0.03836104$$
$$\lambda_2 = 0.27236348$$
$$\lambda_{54} = 16.08465808$$

and at the lower end we have K = 69. We see that the results, shown in fig.7.2, follow the same pattern as in the second example. The results for SOR are not directly comparable to the others since for that run a better starting vector was used. Lanczos needs around 40 and c-g around 60 iterations for full accuracy of the eigenvalue. 62 iterations on 5 vectors gave the 3 lowest eigenvalues when simultaneous iteration was used.

Finally, we consider the one-dimensional biharmonic matrix

$$A = \begin{bmatrix} 5 & -4 & 1 & 0 \dots & 0 \\ -4 & 6 & -4 & 1 \dots & 0 \\ 1 & -4 & 6 & -4 \dots & 0 \\ & & & & \\ 0 & 0 & 0 & 0 \dots & 5 \end{bmatrix}$$

Its eigenvalues

$$\lambda_k = 16 \sin^4 \frac{k\pi}{2(n+1)} \qquad k = 1,\dots,n$$

are densely clustered in the lower end. We have $K = 3 \cdot 10^3$ for n = 20, which means that the power method needs several thousands of iterations to converge.

Now we see from the results in fig. 7.3 that the more sophisticated Lanczos and c-g methods converge much faster than the simpler SOR.

This is due to the fact that this matrix does not, even approximately, behave like those 2-cyclic matrices for which the SOR theory was tailored. Study fig. 7.4a where the eigenvalues of the SOR iteration matrix are plotted for $\omega=1.80$, the value for which the fastest convergence occurred. The closeness of the two eigenvalues near 1 is what ruins the convergence. Compare this to fig. 7.4b where the eigenvalues of the SOR matrix obtained when seeking the eigenvalue in the other end of the spectrum are plotted. Here we see the typical behaviour in well-conditioned cases when the 2-cyclic theory is a good description of the actual performance.

Because the dimension n is only 20, Lanczos' algorithm will converge exceptionally fast since then soon V_s will span the whole space.

Simultaneous iteration needed 119 iterations on 5 vectors to compute the lowest eigenvalue.

Fig. 7.1. Laplace, n=15×20=300

——— Steepest descent — —c-g

—·—SOR ω=1.7 ·····Lanczos

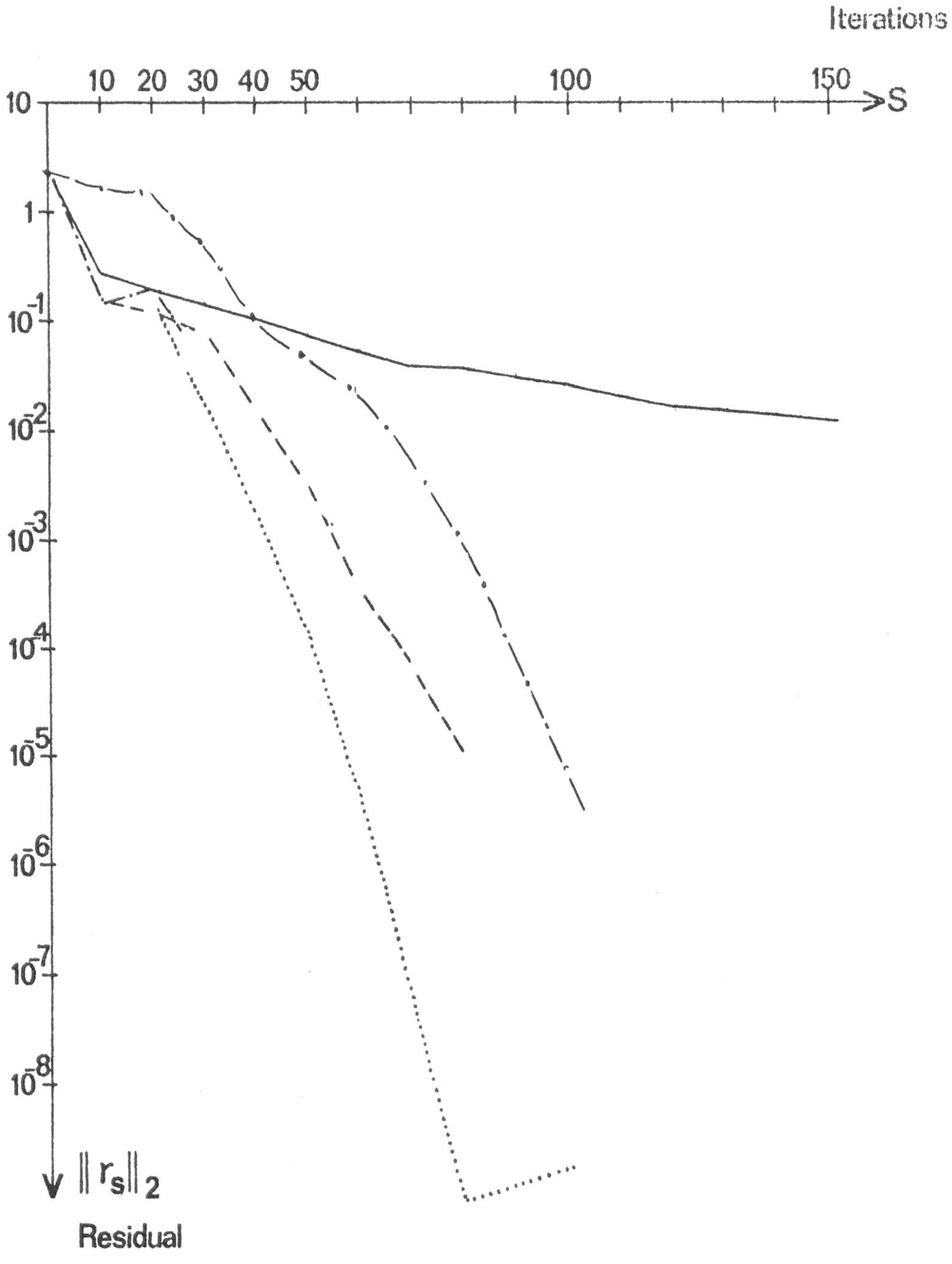

Fig. 7.2. Curtis matrix, n=54

—·—·—SOR ω=1.5 ——— Steepest descent

— — — c-g ······Lanczos

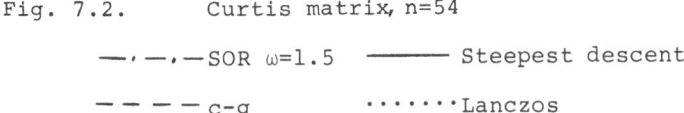

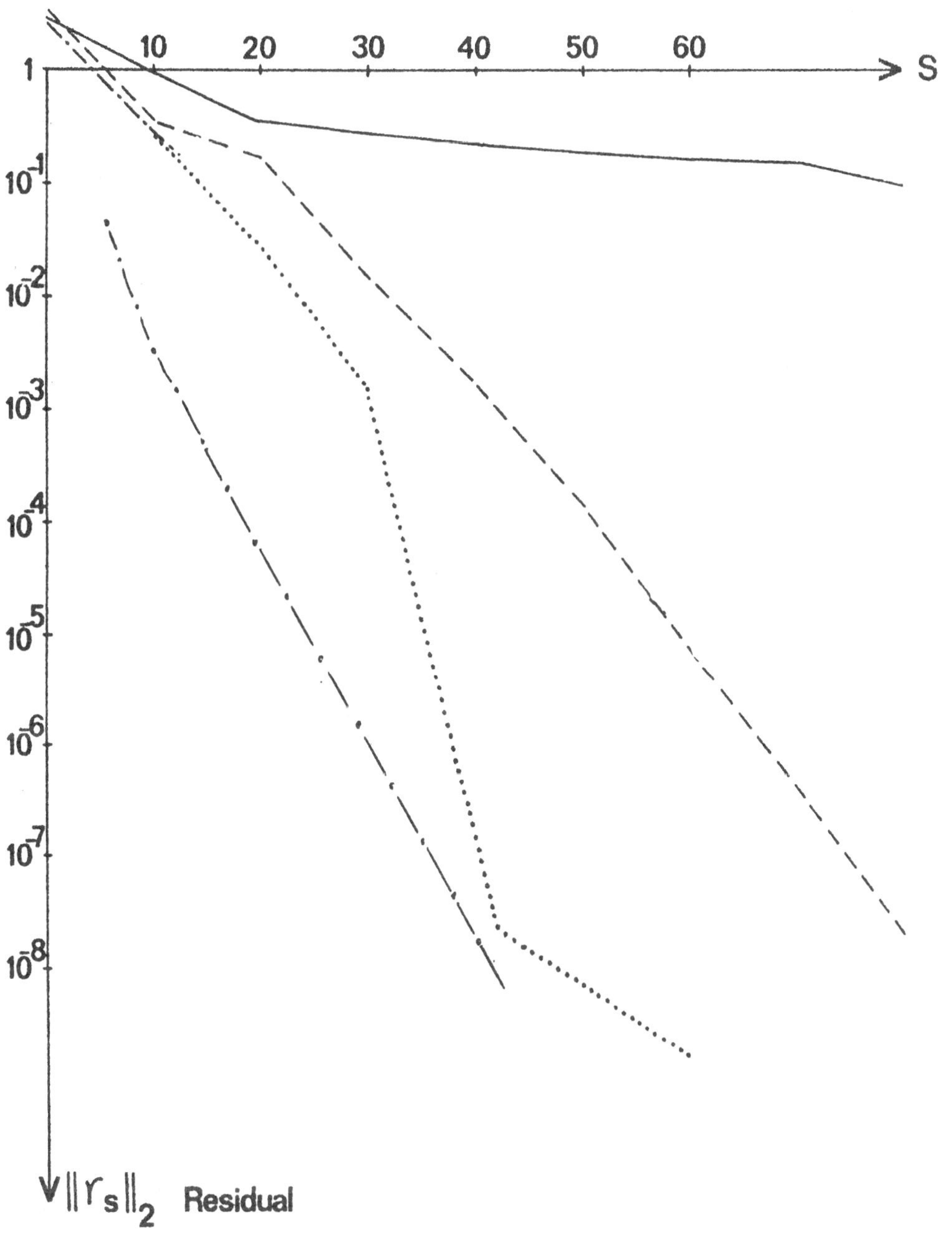

Fig. 7.3. $A=T^2$, smallest eigenvalue, n=20

① Coordinate relaxation ω=1

② ——— " ——— ω=1.80

③ Conjugate gradients

④ Lanczos

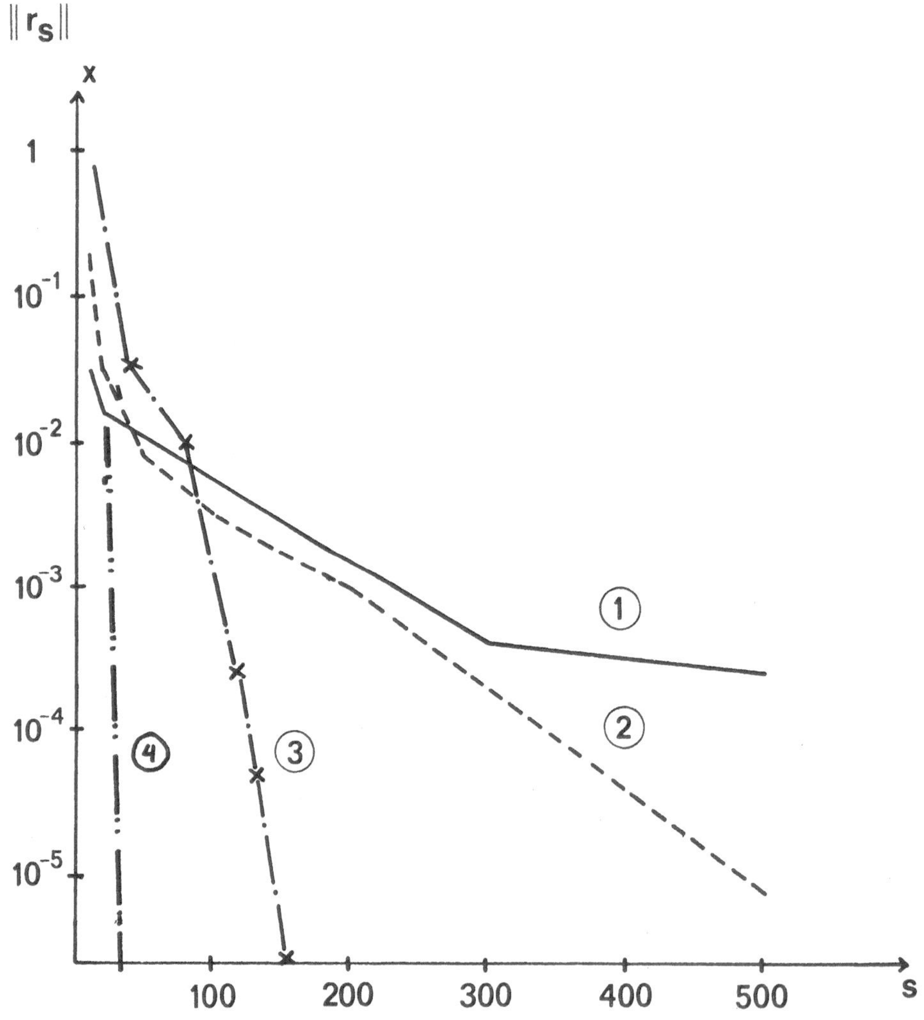

Fig. 7.4a. $A=T^2$, n=20, lower end of spectrum
Eigenvalues of SOR-iteration matrix for ω=1.80 *

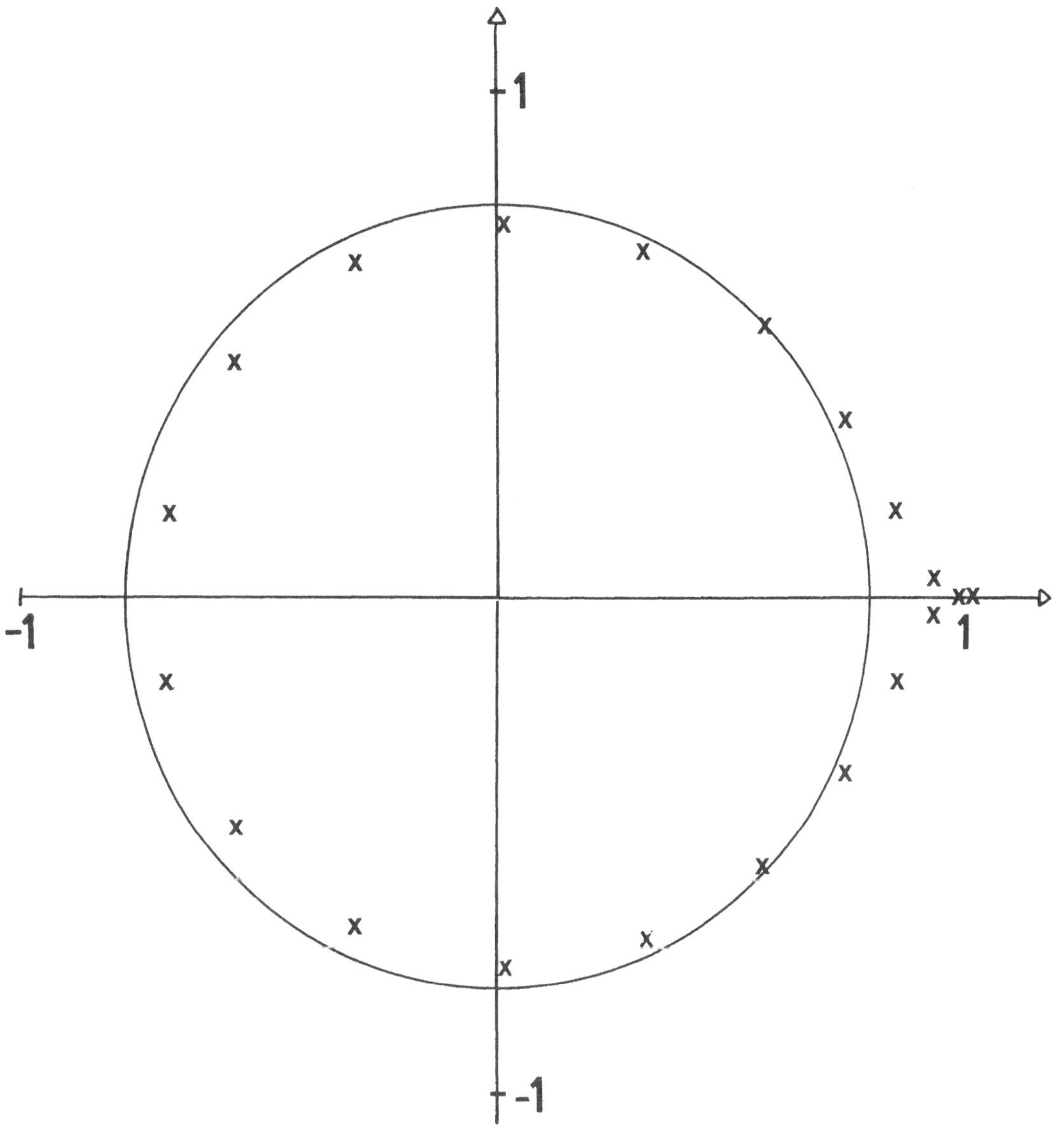

Fig. 7.4b. $A=T^2$, n=20, upper end of spectrum.
Eigenvalues of SOR-iteration matrix for $\omega=1.50(0.05)1.65$*

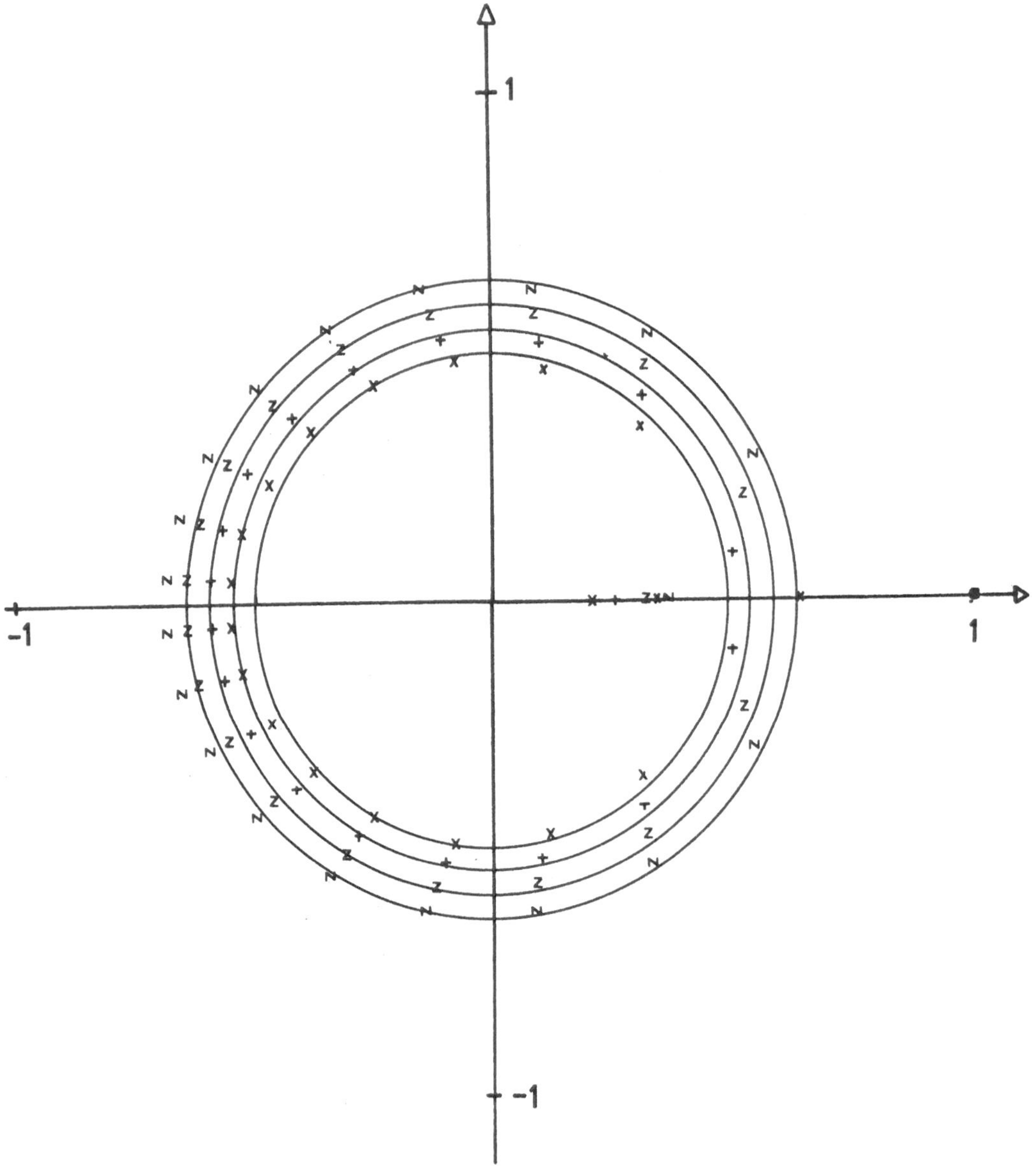

References

1. Axelsson, O.: On preconditioning and convergence acceleration in sparse matrix problems. CERN Data division Tech.Rep. 74-10(1974).

2. Bathe, K-J, and Wilson, E.L.: Solution methods for eigenvalue problems in structural mechanics. Int. J. Num. Meth. Engrg. 6, 213-226(1973).

3. Betteridge, T.: An analytic storage allocation model. Acta Informatica 3, 101-122(1974).

4. Birkhoff, G, and George, A.: Elimination by nested dissection, pp. 221-269 in Complexity of sequential and parallel numerical algorithms, ed. J.F.Traub, Academic Press, (1973).

5. Björck, Å.: Solving linear least squares problems by Gram-Schmidt orthogonalization. BIT 7, 1-21(1967).

6. Bradbury, W.W, and Fletcher, R.: New iterative methods for solution of the eigenproblem. Num. Math. 9, 259-267(1966).

7. Bussemaker, F.C, Cobeljić, S, Cvetković, D.M, Seidel, J.J.: Computer investigation of cubic graphs. Tech.Hog. Eindhoven Rep. 76-WSK-01 (1976).

8. Cline, A.K, Golub, G.H, Platzman, G.W.: Calculation of normal modes of oceans using a Lanczos method, pp 409-426 in Bunch and Rose eds. Sparse matrix comp. Acad. Press (1976).

9. Concus, P, and Golub, G.H.: Use of fast direct methods for the efficient numerical solution of nonseparable elliptic equations. SIAM J. Num. An. 10, 1103-1120(1973).

10. Cullum, J, and Donath, W.E.: A block generalization of the symmetric s-step Lanczos algorithm. IBM Yorktown Heights Tech. Rep. RC 4845 pp. 1-77(1974).

11. Cullum, J, Donath, W.E, Wolfe, P.: The minimization of certain nondifferentiable sums of eigenvalues of symmetric matrices. Math. Progr. Studiy 3, 35-55(1975).

12. Cvetković, D.M.: Graphs and their spectra. Univ. Beograd Publ. Elektr. Fak. Ser. Mat. Fiz 354, 1-50(1971).

13. Duff, I.S, and Reid, J.K.: On the reduction of sparse matrices to condensed forms by similarity transformations. J. Inst. Math. Applics. 15, 217-224(1975).

14. Faddeev, D.K, and Faddeeva, V.N.: Computational methods of linear algebra (Transl.) Freeman & Co., San Fransisco (1963).

15. Fried, I.: Gradient method for finite element eigenproblems. AIAA Jour. 7, 739-741(1969).

16. Fried, I.: Optimal gradient minimization schema for finite element eigenproblems. Jour. Sound and Vibration 20, 333-342(1972).

17. Geradin, M.: The computational efficiency of a new minimization algorithm for eigenvalue analysis. Jour. Sound and Vibration 19, 319-331(1971).

18. Geradin, M.: Analyse dynamique duale des structures par la methode des elements finis. Diss. Univ.de Liege, Belgium, (1972).

19. Handbook for Automatic Computation vol II: Linear Algebra. Wilkinson, J.H, and Reinsch, C. ed. Springer-Verlag, Berlin Heidelberg New York (1971).

20. Householder, A.S.: The theory of matrices in numerical analysis. Blaisdell, New York (1964).

21. Jensen, P.S.: The solution of large symmetric eigenproblems by sectioning. SIAM J. Num. An. 9, 534-545(1972).

22. Kahan, W, and Parlett, B.: An analysis of Lanczos algorithms for symmetric matrices. Tech. Rep. Electronics Research Laboratory, University of California, Berkeley, (1974).

23. Kahan, W, and Parlett, B.N.: How far should you go with the Lanczos process? pp. 131-144 in Bunch and Rose eds. Sparse matrix computations. Acad. Press (1976).

24. Lanczos, C.: An iteration method for the solution of the eigenvalue problem of linear differential and integral operators. NBS J. Res. 45, 255-282(1950).

25. Moler, C.B, and Stewart, G.W.: An algorithm for the generalized matrix eigenvalue problem $Ax = \lambda Bx$. SIAM J. Num. An. 10, 241-256 (1973).

26. Ostrowski, A.M.: On the convergence of the Rayleigh quotient iteration for the computation of characteristic roots and vectors Arch Rat. Mech. Anal. 1, 233-241, 2, 423-428, 3, 325-340, 3, 341-347, 3, 428-481, 4, 153-165(1958).

27. Paige, C.C.: The computation of eigenvalues and eigenvectors of very large sparse matrices. Diss. London Univ. Institute of Computer Science, (1971).

28. Paige, C.C.: Computational variants of the Lanczos method for the eigenproblem. J. Inst. Maths. Applics. 10, 373-381(1972).

29. Paige, C.C, and Saunders, M.A.: Solution of sparse indefinite systems of linear equations. SIAM J. Num. An. Vol. 12, 617-629 (1975).

30. Parlett, B.N, and Kahan, W.M.: On the convergence of a practical QR algorithm. Proc IFIP Congress (1968).

31. Reid, J.K.: Sparse matrices. Tech. Rep. CSS 31 Comp. Sc. Syst. Div. AERE Harwell (1976).

32. Rodrigue, G.: A gradient method for the matrix eigenvalue problem $Ax = \lambda Bx$. Num. Math. 22, 1-16 (1973).

33. Ruhe, A.: Iterative eigenvalue algorithms for large symmetric matrices, ISNM 24 Birkhäuser verlag, Basel und Stuttgart, pp. 97-112(1974).

34. Ruhe, A.: SOR-methods for the eigenvalue problem with large sparse matrices. Math. Comp. 28, 695-710(1974).

35. Ruhe, A.: Iterative eigenvalue algorithms based on convergent splittings. J. Comp. Phys. 19, 110-120(1975).

36. Schwarz, H.R.: The eigenvalue problem $(A-\lambda B)x=0$ for symmetric matrices of high order. Comp. Meth. Appl. Mech. and Engineering 3, 11-28(1974).

37. Shavitt, I, Bender, C.F, Pipano, A, Hosteny, R.P.: The iterative calculation of several of the lowest or highest eigenvalues and corresponding eigenvectors of very large symmetric matrices. Jour. Comp. Phys. 11, 90-108(1973).

38. Stewart, G.W.: The numerical treatment of large eigenvalue problems. pp. 666-672 in IFIP 74-North Holland (1974).

39. Stewart, G.W.: A bibliographical tour of the large sparse generalized eigenvalue problem. pp. 113-130 of Bunch and Rose eds. Sparse matrix computations, Acad. Press (1976).

40. Stewart, W.J.: Markov analysis of operating system techniques. Diss. Queen's Univ. Belfast (1974).

41. Underwood, R.: An iterative block Lanczos method for the solution of large sparse symmetric eigenproblems. Tech. Rep. STAN-CS-75-496 Stanford University (1975).

42. Weaver, W, and Yoshida, D.M.: The eigenvalue problem for banded matrices. Computers & Structures 1, 651-664(1971).

43. Wilkinson, J.H.: The Algebraic Eigenvalue Problem. Clarendon Press, Oxford (1965).

44. Wilkinson, J.H.: Inverse iteration in theory and in practice. Symp. Math. 10, 361-379(1971/72).

45. Young, D.M.: Iterative solution of large linear systems. Academic Press, New York (1971).

Vol. 399: Functional Analysis and its Applications. Proceedings 1973. Edited by H. G. Garnir, K. R. Unni and J. H. Williamson. II, 584 pages. 1974.

Vol. 400: A Crash Course on Kleinian Groups. Proceedings 1974. Edited by L. Bers and I. Kra. VII, 130 pages. 1974.

Vol. 401: M. F. Atiyah, Elliptic Operators and Compact Groups. V, 93 pages. 1974.

Vol. 402: M. Waldschmidt, Nombres Transcendants. VIII, 277 pages. 1974.

Vol. 403: Combinatorial Mathematics. Proceedings 1972. Edited by D. A. Holton. VIII, 148 pages. 1974.

Vol. 404: Théorie du Potentiel et Analyse Harmonique. Edité par J. Faraut. V, 245 pages. 1974.

Vol. 405: K. J. Devlin and H. Johnsbråten, The Souslin Problem. VIII, 132 pages. 1974.

Vol. 406: Graphs and Combinatorics. Proceedings 1973. Edited by R. A. Bari and F. Harary. VIII, 355 pages. 1974.

Vol. 407: P. Berthelot, Cohomologie Cristalline des Schémas de Caracteristique p > o. II, 604 pages. 1974.

Vol. 408: J. Wermer, Potential Theory. VIII, 146 pages. 1974.

Vol. 409: Fonctions de Plusieurs Variables Complexes, Séminaire François Norguet 1970-1973. XIII, 612 pages. 1974.

Vol. 410: Séminaire Pierre Lelong (Analyse) Année 1972-1973. VI, 181 pages. 1974.

Vol. 411: Hypergraph Seminar. Ohio State University, 1972. Edited by C. Berge and D. Ray-Chaudhuri. IX, 287 pages. 1974.

Vol. 412: Classification of Algebraic Varieties and Compact Complex Manifolds. Proceedings 1974. Edited by H. Popp. V, 333 pages. 1974.

Vol. 413: M. Bruneau, Variation Totale d'une Fonction. XIV, 332 pages. 1974.

Vol. 414: T. Kambayashi, M. Miyanishi and M. Takeuchi, Unipotent Algebraic Groups. VI, 165 pages. 1974.

Vol. 415: Ordinary and Partial Differential Equations. Proceedings 1974. XVII, 447 pages. 1974.

Vol. 416: M. E. Taylor, Pseudo Differential Operators. IV, 155 pages. 1974.

Vol. 417: H. H. Keller, Differential Calculus in Locally Convex Spaces. XVI, 131 pages. 1974.

Vol. 418: Localization in Group Theory and Homotopy Theory and Related Topics. Battelle Seattle 1974 Seminar. Edited by P. J. Hilton. VI, 172 pages 1974.

Vol. 419: Topics in Analysis. Proceedings 1970. Edited by O. E. Lehto, I. S. Louhivaara, and R. H. Nevanlinna. XIII, 392 pages. 1974.

Vol. 420: Category Seminar. Proceedings 1972/73. Edited by G. M. Kelly. VI, 375 pages. 1974.

Vol. 421: V. Poénaru, Groupes Discrets. VI, 216 pages. 1974.

Vol. 422: J.-M. Lemaire, Algèbres Connexes et Homologie des Espaces de Lacets. XIV, 133 pages. 1974.

Vol. 423: S. S. Abhyankar and A. M. Sathaye, Geometric Theory of Algebraic Space Curves. XIV, 302 pages. 1974.

Vol. 424: L. Weiss and J. Wolfowitz, Maximum Probability Estimators and Related Topics. V, 106 pages. 1974.

Vol. 425: P. R. Chernoff and J. E. Marsden, Properties of Infinite Dimensional Hamiltonian Systems. IV, 160 pages. 1974.

Vol. 426: M. L. Silverstein, Symmetric Markov Processes. X, 287 pages. 1974.

Vol. 427: H. Omori, Infinite Dimensional Lie Transformation Groups. XII, 149 pages. 1974.

Vol. 428: Algebraic and Geometrical Methods in Topology, Proceedings 1973. Edited by L. F. McAuley. XI, 280 pages. 1974.

Vol. 429: L. Cohn, Analytic Theory of the Harish-Chandra C-Function. III, 154 pages. 1974.

Vol. 430: Constructive and Computational Methods for Differential and Integral Equations. Proceedings 1974. Edited by D. L. Colton and R. P. Gilbert. VII, 476 pages. 1974.

Vol. 431: Séminaire Bourbaki - vol. 1973/74. Exposés 436-452. IV, 347 pages. 1975.

Vol. 432: R. P. Pflug, Holomorphiegebiete, pseudokonvexe Gebiete und das Levi-Problem. VI, 210 Seiten. 1975.

Vol. 433: W. G. Faris, Self-Adjoint Operators. VII, 115 pages. 1975.

Vol. 434: P. Brenner, V. Thomée, and L. B. Wahlbin, Besov Spaces and Applications to Difference Methods for Initial Value Problems. II, 154 pages. 1975.

Vol. 435: C. F. Dunkl and D. E. Ramirez, Representations of Commutative Semitopological Semigroups. VI, 181 pages. 1975.

Vol. 436: L. Auslander and R. Tolimieri, Abelian Harmonic Analysis, Theta Functions and Function Algebras on a Nilmanifold. V, 99 pages. 1975.

Vol. 437: D. W. Masser, Elliptic Functions and Transcendence. XIV, 143 pages. 1975.

Vol. 438: Geometric Topology. Proceedings 1974. Edited by L. C. Glaser and T. B. Rushing. X, 459 pages. 1975.

Vol. 439: K. Ueno, Classification Theory of Algebraic Varieties and Compact Complex Spaces. XIX, 278 pages. 1975

Vol. 440: R. K. Getoor, Markov Processes: Ray Processes and Right Processes. V, 118 pages. 1975.

Vol. 441: N. Jacobson, PI-Algebras. An Introduction. V, 115 pages. 1975.

Vol. 442: C. H. Wilcox, Scattering Theory for the d'Alembert Equation in Exterior Domains. III, 184 pages. 1975.

Vol. 443: M. Lazard, Commutative Formal Groups. II, 236 pages. 1975.

Vol. 444: F. van Oystaeyen, Prime Spectra in Non-Commutative Algebra. V, 128 pages. 1975.

Vol. 445: Model Theory and Topoi. Edited by F. W. Lawvere, C. Maurer, and G. C. Wraith. III, 354 pages. 1975.

Vol. 446: Partial Differential Equations and Related Topics. Proceedings 1974. Edited by J. A. Goldstein. IV, 389 pages. 1975.

Vol. 447: S. Toledo, Tableau Systems for First Order Number Theory and Certain Higher Order Theories. III, 339 pages. 1975.

Vol. 448: Spectral Theory and Differential Equations. Proceedings 1974. Edited by W. N. Everitt. XII, 321 pages. 1975.

Vol. 449: Hyperfunctions and Theoretical Physics. Proceedings 1973. Edited by F. Pham. IV, 218 pages. 1975.

Vol. 450: Algebra and Logic. Proceedings 1974. Edited by J. N. Crossley. VIII, 307 pages. 1975.

Vol. 451: Probabilistic Methods in Differential Equations. Proceedings 1974. Edited by M. A. Pinsky. VII, 162 pages. 1975.

Vol. 452: Combinatorial Mathematics III. Proceedings 1974. Edited by Anne Penfold Street and W. D. Wallis. IX, 233 pages. 1975.

Vol. 453: Logic Colloquium. Symposium on Logic Held at Boston, 1972-73. Edited by R. Parikh. IV, 251 pages. 1975.

Vol. 454: J. Hirschfeld and W. H. Wheeler, Forcing, Arithmetic, Division Rings. VII, 266 pages. 1975.

Vol. 455: H. Kraft, Kommutative algebraische Gruppen und Ringe. III, 163 Seiten. 1975.

Vol. 456: R. M. Fossum, P. A. Griffith, and I. Reiten, Trivial Extensions of Abelian Categories. Homological Algebra of Trivial Extensions of Abelian Categories with Applications to Ring Theory. XI, 122 pages. 1975.

Vol. 457: Fractional Calculus and Its Applications. Proceedings 1974. Edited by B. Ross. VI, 381 pages. 1975.

Vol. 458: P. Walters, Ergodic Theory – Introductory Lectures. VI, 198 pages. 1975.

Vol. 459: Fourier Integral Operators and Partial Differential Equations. Proceedings 1974. Edited by J. Chazarain. VI, 372 pages. 1975.

Vol. 460: O. Loos, Jordan Pairs. XVI, 218 pages. 1975.

Vol. 461: Computational Mechanics. Proceedings 1974. Edited by J. T. Oden. VII, 328 pages. 1975.

Vol. 462: P. Gérardin, Construction de Séries Discrètes p-adiques. »Sur les séries discrètes non ramifiées des groupes réductifs déployés p-adiques«. III, 180 pages. 1975.

Vol. 463: H.-H. Kuo, Gaussian Measures in Banach Spaces. VI, 224 pages. 1975.

Vol. 464: C. Rockland, Hypoellipticity and Eigenvalue Asymptotics. III, 171 pages. 1975.

Vol. 465: Séminaire de Probabilités IX. Proceedings 1973/74. Edité par P. A. Meyer. IV, 589 pages. 1975.

Vol. 466: Non-Commutative Harmonic Analysis. Proceedings 1974. Edited by J. Carmona, J. Dixmier and M. Vergne. VI, 231 pages. 1975.

Vol. 467: M. R. Essén, The Cos $\pi\lambda$ Theorem. With a paper by Christer Borell. VII, 112 pages. 1975.

Vol. 468: Dynamical Systems – Warwick 1974. Proceedings 1973/74. Edited by A. Manning. X, 405 pages. 1975.

Vol. 469: E. Binz, Continuous Convergence on C(X). IX, 140 pages. 1975.

Vol. 470: R. Bowen, Equilibrium States and the Ergodic Theory of Anosov Diffeomorphisms. III, 108 pages. 1975.

Vol. 471: R. S. Hamilton, Harmonic Maps of Manifolds with Boundary. III, 168 pages. 1975.

Vol. 472: Probability-Winter School. Proceedings 1975. Edited by Z. Ciesielski, K. Urbanik, and W. A. Woyczyński. VI, 283 pages. 1975.

Vol. 473: D. Burghelea, R. Lashof, and M. Rothenberg, Groups of Automorphisms of Manifolds. (with an appendix by E. Pedersen) VII, 156 pages. 1975.

Vol. 474: Séminaire Pierre Lelong (Analyse) Année 1973/74. Edité par P. Lelong. VI, 182 pages. 1975.

Vol. 475: Répartition Modulo 1. Actes du Colloque de Marseille-Luminy, 4 au 7 Juin 1974. Edité par G. Rauzy. V, 258 pages. 1975. 1975.

Vol. 476: Modular Functions of One Variable IV. Proceedings 1972. Edited by B. J. Birch and W. Kuyk. V, 151 pages. 1975.

Vol. 477: Optimization and Optimal Control. Proceedings 1974. Edited by R. Bulirsch, W. Oettli, and J. Stoer. VII, 294 pages. 1975.

Vol. 478: G. Schober, Univalent Functions – Selected Topics. V, 200 pages. 1975.

Vol. 479: S. D. Fisher and J. W. Jerome, Minimum Norm Extremals in Function Spaces. With Applications to Classical and Modern Analysis. VIII, 209 pages. 1975.

Vol. 480: X. M. Fernique, J. P. Conze et J. Gani, Ecole d'Eté de Probabilités de Saint-Flour IV–1974. Edité par P.-L. Hennequin. XI, 293 pages. 1975.

Vol. 481: M. de Guzmán, Differentiation of Integrals in R^n. XII, 226 pages. 1975.

Vol. 482: Fonctions de Plusieurs Variables Complexes II. Séminaire François Norguet 1974–1975. IX, 367 pages. 1975.

Vol. 483: R. D. M. Accola, Riemann Surfaces, Theta Functions, and Abelian Automorphisms Groups. III, 105 pages. 1975.

Vol. 484: Differential Topology and Geometry. Proceedings 1974. Edited by G. P. Joubert, R. P. Moussu, and R. H. Roussarie. IX, 287 pages. 1975.

Vol. 485: J. Diestel, Geometry of Banach Spaces – Selected Topics. XI, 282 pages. 1975.

Vol. 486: S. Stratila and D. Voiculescu, Representations of AF-Algebras and of the Group U (∞). IX, 169 pages. 1975.

Vol. 487: H. M. Reimann und T. Rychener, Funktionen beschränkter mittlerer Oszillation. VI, 141 Seiten. 1975.

Vol. 488: Representations of Algebras, Ottawa 1974. Proceedings 1974. Edited by V. Dlab and P. Gabriel. XII, 378 pages. 1975.

Vol. 489: J. Bair and R. Fourneau, Etude Géométrique des Espaces Vectoriels. Une Introduction. VII, 185 pages. 1975.

Vol. 490: The Geometry of Metric and Linear Spaces. Proceedings 1974. Edited by L. M. Kelly. X, 244 pages. 1975.

Vol. 491: K. A. Broughan, Invariants for Real-Generated Uniform Topological and Algebraic Categories. X, 197 pages. 1975.

Vol. 492: Infinitary Logic: In Memoriam Carol Karp. Edited by D. W. Kueker. VI, 206 pages. 1975.

Vol. 493: F. W. Kamber and P. Tondeur, Foliated Bundles and Characteristic Classes. XIII, 208 pages. 1975.

Vol. 494: A Cornea and G. Licea. Order and Potential Resolvent Families of Kernels. IV, 154 pages. 1975.

Vol. 495: A. Kerber, Representations of Permutation Groups II. V, 175 pages. 1975.

Vol. 496: L. H. Hodgkin and V. P. Snaith, Topics in K-Theory. Two Independent Contributions. III, 294 pages. 1975.

Vol. 497: Analyse Harmonique sur les Groupes de Lie. Proceedings 1973–75. Edité par P. Eymard et al. VI, 710 pages. 1975.

Vol. 498: Model Theory and Algebra. A Memorial Tribute to Abraham Robinson. Edited by D. H. Saracino and V. B. Weispfenning. X, 463 pages. 1975.

Vol. 499: Logic Conference, Kiel 1974. Proceedings. Edited by G. H. Müller, A. Oberschelp, and K. Potthoff. V, 651 pages 1975.

Vol. 500: Proof Theory Symposion, Kiel 1974. Proceedings. Edited by J. Diller and G. H. Müller. VIII, 383 pages. 1975.

Vol. 501: Spline Functions, Karlsruhe 1975. Proceedings. Edited by K. Böhmer, G. Meinardus, and W. Schempp. VI, 421 pages. 1976.

Vol. 502: János Galambos, Representations of Real Numbers by Infinite Series. VI, 146 pages. 1976.

Vol. 503: Applications of Methods of Functional Analysis to Problems in Mechanics. Proceedings 1975. Edited by P. Germain and B. Nayroles. XIX, 531 pages. 1976.

Vol. 504: S. Lang and H. F. Trotter, Frobenius Distributions in GL_2-Extensions. III, 274 pages. 1976.

Vol. 505: Advances in Complex Function Theory. Proceedings 1973/74. Edited by W. E. Kirwan and L. Zalcman. VIII, 203 pages. 1976.

Vol. 506: Numerical Analysis, Dundee 1975. Proceedings. Edited by G. A. Watson. X, 201 pages. 1976.

Vol. 507: M. C. Reed, Abstract Non-Linear Wave Equations. VI, 128 pages. 1976.

Vol. 508: E. Seneta, Regularly Varying Functions. V, 112 pages. 1976.

Vol. 509: D. E. Blair, Contact Manifolds in Riemannian Geometry. VI, 146 pages. 1976.

Vol. 510: V. Poènaru, Singularités C^∞ en Présence de Symétrie. V, 174 pages. 1976.

Vol. 511: Séminaire de Probabilités X. Proceedings 1974/75. Edité par P. A. Meyer. VI, 593 pages. 1976.

Vol. 512: Spaces of Analytic Functions, Kristiansand, Norway 1975. Proceedings. Edited by O. B. Bekken, B. K. Øksendal, and A. Stray. VIII, 204 pages. 1976.

Vol. 513: R. B. Warfield, Jr. Nilpotent Groups. VIII, 115 pages. 1976.

Vol. 514: Séminaire Bourbaki vol. 1974/75. Exposés 453 – 470. IV, 276 pages. 1976.

Vol. 515: Bäcklund Transformations. Nashville, Tennessee 1974. Proceedings. Edited by R. M. Miura. VIII, 295 pages. 1976.

Vol. 516: M. L. Silverstein, Boundary Theory for Symmetric Markov Processes. XVI, 314 pages. 1976.

Vol. 517: S. Glasner, Proximal Flows. VIII, 153 pages. 1976.

Vol. 518: Séminaire de Théorie du Potentiel, Proceedings Paris 1972–1974. Edité par F. Hirsch et G. Mokobodzki. VI, 275 pages. 1976.

Vol. 519: J. Schmets, Espaces de Fonctions Continues. XII, 150 pages. 1976.

Vol. 520: R. H. Farrell, Techniques of Multivariate Calculation. X, 337 pages. 1976.

Vol. 521: G. Cherlin, Model Theoretic Algebra – Selected Topics. IV, 234 pages. 1976.

Vol. 522: C. O. Bloom and N. D. Kazarinoff, Short Wave Radiation Problems in Inhomogeneous Media: Asymptotic Solutions. V. 104 pages. 1976.

Vol. 523: S. A. Albeverio and R. J. Høegh-Krohn, Mathematical Theory of Feynman Path Integrals. IV, 139 pages. 1976.

Vol. 524: Séminaire Pierre Lelong (Analyse) Année 1974/75. Edité par P. Lelong. V, 222 pages. 1976.

Vol. 525: Structural Stability, the Theory of Catastrophes, and Applications in the Sciences. Proceedings 1975. Edited by P. Hilton. VI, 408 pages. 1976.

Vol. 526: Probability in Banach Spaces. Proceedings 1975. Edited by A. Beck. VI, 290 pages. 1976.

Vol. 527: M. Denker, Ch. Grillenberger, and K. Sigmund, Ergodic Theory on Compact Spaces. IV, 360 pages. 1976.

Vol. 528: J. E. Humphreys, Ordinary and Modular Representations of Chevalley Groups. III, 127 pages. 1976.

Vol. 529: J. Grandell, Doubly Stochastic Poisson Processes. X, 234 pages. 1976.

Vol. 530: S. S. Gelbart, Weil's Representation and the Spectrum of the Metaplectic Group. VII, 140 pages. 1976.

Vol. 531: Y.-C. Wong, The Topology of Uniform Convergence on Order-Bounded Sets. VI, 163 pages. 1976.

Vol. 532: Théorie Ergodique. Proceedings 1973/1974. Edité par J.-P. Conze and M. S. Keane. VIII, 227 pages. 1976.

Vol. 533: F. R. Cohen, T. J. Lada, and J. P. May, The Homology of Iterated Loop Spaces. IX, 490 pages. 1976.

Vol. 534: C. Preston, Random Fields. V, 200 pages. 1976.

Vol. 535: Singularités d'Applications Differentiables. Plans-sur-Bex. 1975. Edité par O. Burlet et F. Ronga. V, 253 pages. 1976.

Vol. 536: W. M. Schmidt, Equations over Finite Fields. An Elementary Approach. IX, 267 pages. 1976.

Vol. 537: Set Theory and Hierarchy Theory. Bierutowice, Poland 1975. A Memorial Tribute to Andrzej Mostowski. Edited by W. Marek, M. Srebrny and A. Zarach. XIII, 345 pages. 1976.

Vol. 538: G. Fischer, Complex Analytic Geometry. VII, 201 pages. 1976.

Vol. 539: A. Badrikian, J. F. C. Kingman et J. Kuelbs, Ecole d'Eté de Probabilités de Saint Flour V-1975. Edité par P.-L. Hennequin. IX, 314 pages. 1976.

Vol. 540: Categorical Topology, Proceedings 1975. Edited by E. Binz and H. Herrlich. XV, 719 pages. 1976.

Vol. 541: Measure Theory, Oberwolfach 1975. Proceedings. Edited by A. Bellow and D. Kölzow. XIV, 430 pages. 1976.

Vol. 542: D. A. Edwards and H. M. Hastings, Čech and Steenrod Homotopy Theories with Applications to Geometric Topology. VII, 296 pages. 1976.

Vol. 543: Nonlinear Operators and the Calculus of Variations, Bruxelles 1975. Edited by J. P. Gossez, E. J. Lami Dozo, J. Mawhin, and L. Waelbroeck, VII, 237 pages. 1976.

Vol. 544: Robert P. Langlands, On the Functional Equations Satisfied by Eisenstein Series. VII, 337 pages. 1976.

Vol. 545: Noncommutative Ring Theory. Kent State 1975. Edited by J. H. Cozzens and F. L. Sandomierski. V, 212 pages. 1976.

Vol. 546: K. Mahler, Lectures on Transcendental Numbers. Edited and Completed by B. Diviš and W. J. Le Veque. XXI, 254 pages. 1976.

Vol. 547: A. Mukherjea and N. A. Tserpes, Measures on Topologica Semigroups: Convolution Products and Random Walks. V, 197 pages. 1976.

Vol. 548: D. A. Hejhal, The Selberg Trace Formula for PSL (2,ℝ) Volume I. VI, 516 pages. 1976.

Vol. 549: Brauer Groups, Evanston 1975. Proceedings. Edited b D. Zelinsky. V, 187 pages. 1976.

Vol. 550: Proceedings of the Third Japan – USSR Symposium o Probability Theory. Edited by G. Maruyama and J. V. Prokhorov. V 722 pages. 1976.

Vol. 551: Algebraic K-Theory, Evanston 1976. Proceedings. Edite by M. R. Stein. XI, 409 pages. 1976.

Vol. 552: C. G. Gibson, K. Wirthmüller, A. A. du Plessis an E. J. N. Looijenga. Topological Stability of Smooth Mappings. V 155 pages. 1976.

Vol. 553: M. Petrich, Categories of Algebraic Systems. Vector an Projective Spaces, Semigroups, Rings and Lattices. VIII, 217 pages 1976.

Vol. 554: J. D. H. Smith, Mal'cev Varieties. VIII, 158 pages. 1976

Vol. 555: M. Ishida, The Genus Fields of Algebraic Number Fields VII, 116 pages. 1976.

Vol. 556: Approximation Theory. Bonn 1976. Proceedings. Edited b R. Schaback and K. Scherer. VII, 466 pages. 1976.

Vol. 557: W. Iberkleid and T. Petrie, Smooth S^1 Manifolds. III 163 pages. 1976.

Vol. 558: B. Weisfeiler, On Construction and Identification of Graphs XIV, 237 pages. 1976.

Vol. 559: J.-P. Caubet, Le Mouvement Brownien Relativiste. IX 212 pages. 1976.

Vol. 560: Combinatorial Mathematics, IV, Proceedings 1975. Edite by L. R. A. Casse and W. D. Wallis. VII, 249 pages. 1976.

Vol. 561: Function Theoretic Methods for Partial Differential Equations Darmstadt 1976. Proceedings. Edited by V. E. Meister, N. Wecl and W. L. Wendland. XVIII, 520 pages. 1976.

Vol. 562: R. W. Goodman, Nilpotent Lie Groups: Structure an Applications to Analysis. X, 210 pages. 1976.

Vol. 563: Séminaire de Théorie du Potentiel. Paris, No. 2. Proceedings 1975–1976. Edited by F. Hirsch and G. Mokobodzki. VI, 292 pages 1976.

Vol. 564: Ordinary and Partial Differential Equations, Dundee 1976 Proceedings. Edited by W. N. Everitt and B. D. Sleeman. XVIII, 55 pages. 1976.

Vol. 565: Turbulence and Navier Stokes Equations. Proceedings 1975. Edited by R. Temam. IX, 194 pages. 1976.

Vol. 566: Empirical Distributions and Processes. Oberwolfach 1976 Proceedings. Edited by P. Gaenssler and P. Révész. VII, 146 pages 1976.

Vol. 567: Séminaire Bourbaki vol. 1975/76. Exposés 471–488. IV, 303 pages. 1977.

Vol. 568: R. E. Gaines and J. L. M. Mawhin, Alternative Problems, Coincidence Degree, and Nonlinear Differential Equations. V, 262 pages. 1977.

Vol. 569: Cohomologie Etale SGA 4½. Séminaire de Géométrie Algébrique du Bois-Marie. Edité par P. Deligne. V, 312 pages. 1977.

Vol. 570: Differential Geometrical Methods in Mathematical Physics, Bonn 1975. Proceedings. Edited by K. Bleuler and A. Reetz. VIII, 576 pages. 1977.

Vol. 571: Constructive Theory of Functions of Several Variables, Oberwohlfach 1976. Proceedings. Edited by W. Schempp and K. Zeller. VI, 290 pages. 1977.

Vol. 572: Sparse Matrix Techniques, Copenhagen 1976. Edited by V. A. Barker. V, 184 pages. 1977.